stars + planets
AN ILLUSTRATED GUIDE

This is a Star Fire Book
First published in 2002
This new edition published 2007

07 09 11 10 08

1 3 5 7 9 10 8 6 4 2

Star Fire is part of
The Foundry Creative Media Company Limited
Crabtree Hall, Crabtree Lane, Fulham, London, SW6 6TY

www.star-fire.co.uk

Copyright © 2007 The Foundry

ISBN-13 978 1 84451 461 8
ISBN-10 1 84451 461 7

A copy of the CIP data for this book is available from the British Library

Printed in China

SPECIAL THANKS TO EVERYONE INVOLVED WITH THIS PROJECT:
This edition: Catherine Emslie, Mike Spender, Claire Walker.
Original edition: Lisa Balkwill, Jennifer Bishop, Penny Brown, Antony Cohen, Duncan Copp, Sarah Goulding,
Ian Graham, Chris Herbert, Ann Kay, Sonya Newland, Colin Rudderham, Graham Stride, Nick Wells, Polly Willis.

stars + planets
AN ILLUSTRATED GUIDE

Chris Cooper, Pam Spence, Carole Stott

GENERAL EDITOR: IAIN NICOLSON

STAR FIRE

CONTENTS

INTRODUCTION

Thanks to the efforts of astronomers over the ages, we now have a comprehensive understanding of the Universe in which we live. We know that the Earth is one of nine planets that revolve around the Sun and that the Sun itself is an ordinary star. As such, it is an incandescent globe of high-temperature gas, powered by nuclear reactions taking place deep within its core. Each of the member planets of our Solar System (the system of bodies that revolves around the Sun) is a fascinating world in its own right. Some, such as Venus or Mars, share some common characteristics with the Earth, others, such as Jupiter and Saturn, are of a wholly different nature.

The stars are themselves suns, powered by nuclear furnaces. Some are thousands, even millions of times more luminous than our Sun, whereas others are cosmic glow-worms, thousands, or tens of thousands of times less brilliant than our neighbourhood star. Stars appear as mere points of light simply because they are so much further away from us than the Sun, so far, indeed, that light – travelling at 300,000 kilometres (186,000 miles) per second – takes more than four years to reach us from the nearest of them.

Astronomers have mapped out the life cycles of stars from birth to death and have come to realise that the very elements of which we, and planet Earth alike, are composed have been synthesised in, and spewed forth from, the interiors of stars. They have discovered a host of intriguing objects such as white dwarfs, neutron stars and, most enigmatic of all, black holes. During the past few years, they have begun to detect planets around other stars (exoplanets). By studying planetary systems beyond our own, we will gain a better understanding of how our own Solar System came into being and should get closer to answering that most tantalising of questions, 'Are we alone?'

The Sun is one of more than a hundred billion stars which, together with clouds of gas and dust, comprise the immense star system that is our Galaxy. Our Galaxy, vast though it is, is merely one of the hundreds of billions of galaxies that populate the observable Universe. Galaxies themselves are clumped together into groups or clusters which

themselves are gathered into loose, straggly structures called superclusters. Each galaxy, or cluster of galaxies, is receding from every other one; the entire Universe is expanding. Current data suggests that the Universe originated 12–15 billion years ago by erupting forth from an initial state of inconceivably high density and temperature – an event that has come to be known as the Big Bang.

By studying distant galaxies – some so remote that their light has taken ten billion years or more to reach us – and by analyzing the faint background of primordial radiation that permeates all of space, astronomers can investigate the processes that gave birth to galaxies and the first generation of stars, determine the overall 'shape' of space and chart the future evolution of the Universe. They have discovered that the Universe contains far more dark matter than luminous matter and that it may be dominated by a mysterious form of 'dark energy' that is causing its expansion to accelerate rather than decelerate.

In addition to sophisticated ground-based telescopes, astronomers have access to instruments borne aloft on high-altitude balloons and transported above the obscuring effects of our atmosphere aboard orbiting satellites. As a result, distant stars and galaxies can now be studied across virtually the entire electromagnetic spectrum, from gamma- and X-rays to radio waves. Closer to home, robot spacecraft have transformed our understanding of Earth's planetary siblings. Astronauts have visited the Moon and have begun to live and work in orbit around the Earth.

This book contains a wealth of information that has been arranged in a convenient thematic format. The reader-friendly text, which is supported by some 300 illustrations, covers a wide range of topics relating to stars and planets and places them in the wider context of the Universe as a whole. Entries dealing with the exploration of space, the history of astronomy and some of the key people who have contributed to the advance of astronomical knowledge, reveal how we have arrived at our present-day view of the Universe.

Aimed at the general reader, who can look up facts in their relevant sections or simply dip into the text at random, this book has sufficient depth and detail to act as an authoritative reference source. With the aid of cross references, the reader can establish links between related topics and build up a broader and deeper understanding of the subject. The reference aspect is reinforced by tables of planetary, stellar and galaxy data, information about human spaceflight missions and an extensive list of relevant web sites.

New developments are taking place all the time in this rapidly-advancing field. In the future, our best contemporary theories will seem simplistic and inadequate. But, for the moment, this book provides a comprehensive guide to our present-day knowledge of stars, planets, and the Universe which, hopefully, will inspire at least some of its readers to delve deeper into this most fascinating and dynamic of subjects.

IAIN NICOLSON

HISTORY OF ASTRONOMY

WORLD ASTRONOMY

BABYLONIAN ASTRONOMY

The Babylonians were the first people to make regular records of celestial phenomena. The basic constellation figures probably date from before 3000 BC, and Mesopotamian tablets from 1100 BC depicting the Sun, Moon and zodiac figures still exist. Dated observations of eclipses, and of the positions of Venus and other planets in relation to the Sun, were recorded on clay tablets. These documents have proved invaluable to modern astronomers. Indeed, the Babylonians of around 1000 BC were the founders of systematic astronomy. They mistakenly believed that there were 360 days in the year, and it may have been in connection with this that they divided the circle into 360 degrees. Their astronomical records were built up for the purposes of political and religious astrology; in ancient times, no one looked at the sky for what we would now call scientific purposes – they saw it instead as a tool for prophecy. While war, famine and disease made human life precarious, the heavens possessed a regularity which seemed to influence the Earth, through the solar seasons or lunar phases. But Babylonian astrology resulted in the basic concepts of scientific astronomy: observational data, mathematical analysis and verifiable predictions.

HINDU ASTRONOMY

While Indian culture itself dates from earlier than 2000 BC, the basic elements of its astronomy are similar to those of Egypt and Mesopotamia. Before c. 1000 BC, the Hindus worked with a 360-day year. The great developments in Hindu astronomy, however, took place at the same time as those of the Babylonians, and it is quite likely that Babylonian astronomical techniques were conveyed to India along trade routes. These techniques included intercalation (the insertion of a period in the calendar to harmonize it with the solar year) and planetary table-making. Following the development of Greek planetary theory, Indian astronomers began to make use of epicycles (circles whose centres move around

ABOVE: A shepherd contemplates the vast expanse of space with stars and planets in ancient times.

the circumference of greater circles) in solving problems involving variable planetary speeds. The sophistication of Hindu astronomy by 300 BC is evidenced by the quality of their lunar tables and by parallel developments in mathematics, which included the form of notation subsequently called 'Arabic' numerals. By this time, the modified Egyptian 365-day year had come into use, along with a properly intercalated calendrical cycle. By the Hellenistic, Roman and early Christian periods, trade began to flourish between the Mediterranean and India, and texts of Greek and Babylonian authorship began to be translated into Sanskrit.

CHINESE ASTRONOMY

China's astronomical culture dates back to 2,000 BC or earlier. It was the dominant astronomical culture of Japan and Korea also. Chinese astronomy grew

from its own roots, and while wrestling with many of the same problems that astronomers of other traditions confronted – such as reconciling the lunar and solar calendars – it developed independently from the cultures of India and the Middle East. In China, astronomy and politics were intimately connected, as it was an emperor's duty to organize Earthly affairs in conformity with the state of the heavens. This meant that from a very early date official sky-watchers were employed to record the appearances of comets or novae ('new stars'), while the ability to predict eclipses was of paramount importance to political stability. Chinese astronomers recorded a 'guest star' in AD 1054, which we now know to have been the explosion of the supernova whose remains are now visible as the Crab Nebula. The Chinese accurately determined the year to be 365¼ days, and divided the celestial equator into 365¼ degrees. This might be the reason that the Chinese never developed the flexible geometry that would be facilitated by a circle of 360°.

)))➡ **The 360° Circle, p15**

GREEK ASTRONOMY

In the hands of the ancient Greeks, astronomy took great strides as a rational science. In 585 BC Thales is said to have successfully predicted an eclipse, and by 480 BC, Parmenides and others were speaking of a spherical Earth. By 430 BC, the Athenian Meton had elucidated the 19-year cycle of solar eclipses. Eudoxus (407–355 BC) tried to explain the complex motions of the planets with a purely geometric model which contained 27 nested rotating spheres. By 280 BC Aristarchus (c. 310 BC–c. 230 BC) had devised a Sun-centred system, though it could not prevail against the physics of Aristotle (384–322 BC). Aristotle argued that the heavens (which began at the sphere of the Moon) were changeless; comets and new stars must be atmospheric phenomena.

Greek astronomy came to its zenith when it was infused with Babylonian computation techniques and planetary tables in the time of Hipparchus (fl. 160–125 BC). Hipparchus calculated that the Moon's maximum distance was 67.5 earth radii, astonishingly close to the true value of 63.8. He is said to have discovered the precession of the equinoxes and he further developed the system of epicycles of Eudoxus. The system was still further elaborated in the second century AD by the greatest figure of Greek astronomy, Claudius Ptolemy of Alexandria. His geocentric system went unchallenged for 1,400 years.

)))➡ **Epicycle, p16; Ptolemaic System, p16; Hipparchus, p24**

ABOVE: In the eighth century AD, astronomical treatises began to be translated into Arabic from Greek and Sanscrit and other languages.

ARABIC ASTRONOMY

 Islam was founded in AD 622. After the building of a new Muslim capital at Baghdad in AD 762, the Abbasid Caliphs established the House of Wisdom in the new city. Here, Greek and Hindu works on astronomy, medicine and other sciences were translated into Arabic. Astronomy was vital to Muslims: the Qur'an encouraged the study of God's creation; astronomy was necessary to establish prayer times and to find the direction of Mecca, which Muslims must face while praying; and Islam needed a reliable calendar.

One of the greatest of Muslim scientists, Alhazen (c. AD 965–1039), built on Aristotle's *Meteorologia* to do fundamental work in optics, with relevance to astronomy. One of his most influential studies concerned the refraction of light by the atmosphere. Why, for example, do the Sun and Moon look distorted on the horizon, and why is there twilight before sunrise and after sunset?

Arabic astronomy was closely related to the other sciences, such as medicine. Human health was believed to be sensitive to the planets and their movements, and Arabic physicians therefore made use of astrology. At the same time Arabic astronomy developed a strong observational tradition. Al-Sufi produced the first major Islamic star chart c. 964 in Baghdad. Notable observatories included those at Maraghah in Iran, and at Samarkand, in Uzbekistan.

➠ *Babylonian Astronomy, Chinese Astronomy, p8*

ALCHEMY

In Arabic science, astronomy was closely related to alchemy, the forerunner of scientific chemistry. Jabir ibn Hayyan, who died c. AD 803, devised distillation techniques, which led to the discovery of new substances such as nitric acid and alcohol (in Arabic, *al-kohol),* derived from grape juice. Medieval European alchemy was built upon Arabic foundations, as was metallurgical chemistry. It was believed that the Sun, Moon and planets formed gold, silver and other metals in the Earth; alchemists believed they could speed up this process in their laboratories.

MEDIEVAL ASTRONOMY

In medieval Europe some of the ideas of the Greek scientists survived in Latin poetic digests or encyclopedic works, most notably by Macrobius (c. AD 400), Cassiodorus (c. AD 550) and Boethius (AD 480–524). The most important astronomical writer of this period lived in England: the Venerable Bede (AD 675–735) of Northumbria, the first known English astronomer.

After the Crusades had brought Western Christians into contact with the Arab world there was a twelfth-century revival of physics, philosophy, medicine,

ABOVE RIGHT: By Medieval times, observational astronomy was becoming a more defined science, sparking a revolution in astronomical knowledge.

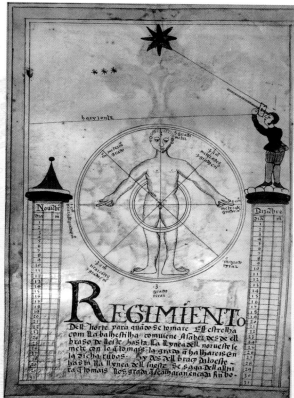

ITALIAN RENAISSANCE

The origins of the sixteenth-century Copernican revolution lie less in errors found in practical astronomy than in the growth of Renaissance humanism. There was an influx of scholars and original Greek manuscripts into Europe following the Turkish occupation of Constantinople in 1453. This meant that European scholars now had access to Ptolemy in the original Greek, rather than through Arabic-to-Latin translations. The Constantinople-educated cardinal and scholar Johannes Bessarion was a major driving force in this movement, and his collection of Greek manuscripts was important in enabling Georg Peurbach (1423–61) and Regiomontanus (1436–76) to produce the *Epitome*, or abridgement, of Ptolemy's *Almagest* (published 1496). It was these accurate Greek interpretations of Ptolemy which truly started the astronomical Renaissance.

))))▶ *Nicolaus Copernicus, p25*

architecture and astronomy. Latin translations of Ptolemy's *Almagest* and Aristotle's *Physics* brought classical Greek science into northern Europe for the first time. Around 1460 Regiomontanus (Johannes Müller, 1436–76) began to measure celestial angles with a set of Ptolemy's Rulers, three graduated rods around 2 m (6 ft) long. Müller's colleague Bernhard Walther compiled the first long run of accurate original positional observations by a north European, between 1475 and 1504.

Following the establishment of Europe's great universities – Bologna, Paris, Oxford, Padua and Montpellier – astronomy became part of the curriculum. By the last quarter of the thirteenth century the scholars of Paris had begun to discuss the possibility of other worlds, while around 1370 the French scientist-bishop Nicole de Oresme suggested the concept that the Earth itself might be spinning on its axis.

))))▶ *Greek Astronomy, p9; Arabic Astronomy, p10*

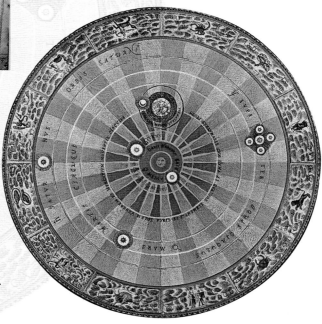

ABOVE: The Copernican Universe, showing the Sun at the centre encircled by Earth and other planets.

INSTRUMENTS AND TOOLS

ARMILLARY SPHERE

Model of the Earth and celestial sphere, first used by Greek astronomers in the second century BC or earlier. A ball representing Earth was circled by rings representing the celestial equator, ecliptic and other reference circles, and the paths of the Sun, Moon and planets.

)))▶ *Astrolabe, below; Celestial Sphere, p167*

ASTROLABE

The astrolabe was an observing and computing instrument that was used in the Middle Ages to determine the time, find direction and make calculations in spherical trigonometry. Its earliest form was described by Ptolemy in about AD 125, but it was the Arabs who developed and popularized the instrument. In its mature form it consisted of a star map in brass filigree, called a rete (network), that rotated above a zonal or 'climate' plate. The plate was engraved with circles representing astronomical coordinate lines appropriate for a particular latitude. Turning the rete simulated the rotation of the stars.

QUADRANT

An astronomical instrument for measuring the altitude of a celestial object (its angular height above the horizon). It had a scale graduated from 0° to 90°. Used since ancient times, it was greatly developed by the Danish observer Tycho Brahe (1546–1601).

)))▶ *Tycho Brahe, p26*

SPECTROSCOPE

A device that analyses light by splitting it into its component colours, using a prism or diffraction grating. In 1666, Isaac Newton (1642–1727) split sunlight into its spectrum, or band of component colours, using a prism. In 1802 William Wollaston (1766–1828) found a few inexplicable black lines in the spectrum of sunlight,

LEFT: The astrolabe was perfected by the medieval Arabs. Its star map and astronomical scales enabled it to be used in navigation and time-finding.

each line representing the absence of a definite colour. In 1814–15 Joseph von Fraunhofer (1787–1826) mapped the positions of 574 black lines seen in sunlight. Such lines in astronomical spectra reveal the identities of atoms and molecules in the bodies that are the sources of the light.

SAMARKAND OBSERVATORY

In 1420 Ulugh Beg (1394–1449), Mongol prince of Samarkand, in what is now Uzbekistan, built a three-storey drum-shaped observatory. It was 30 m (100 ft) high, and housed an enormous masonry sextant of 40-m (130-ft) radius. With this and other instruments at Samarkand, Ulugh Beg determined the length of the solar year to within a minute and compiled a table which gave the precise positions of 1,018 stars.

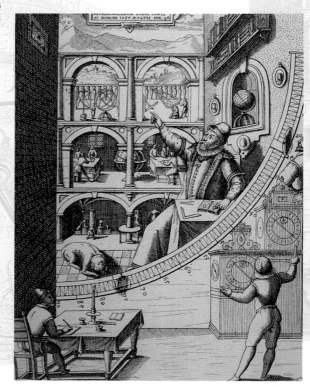

URANIBORG OBSERVATORY

Astronomical observatory on the island of Ven, founded by the Danish astronomer Tycho Brahe (1546–1601) in 1576. Here Tycho developed a succession of large and accurate instruments, including quadrants, sextants and armillary spheres. Uraniborg had its own craftsmen and workshops, so that instrument scales and observing procedures could be constantly monitored and improved. Tycho drew up new tables of planetary motion, and his instruments and observing procedures changed the direction of European astronomy.

LEFT: The observatory at Uraniborg, where Tycho Brahe developed and improved a series of astronomical instruments, including sextants, quadrants and armillary spheres.

ABOVE: The famous Royal Observatory at Greenwich.

GREENWICH OBSERVATORY

The Royal Observatory, Greenwich, was founded in 1675, in the hope that an astronomical solution to the problem of finding longitude at sea could be found. The Reverend John Flamsteed became the first Astronomer Royal at Greenwich, though he had to provide the clocks and instruments from his own resources. Between 1675 and his death in 1719, Flamsteed began an entirely new catalogue of the northern heavens. He was a meticulous observer, and devised and established working procedures for the observatory that have survived to the present day.

CALENDARS

Astrology is the art and science of relating Earthly events and human traits to the movements of the Sun, Moon and planets. It had appeared in Mesopotamia by 2000 BC and passed from there to India, China and Greece. As astronomical tables were compiled over increasing lengths of time, and knowledge of celestial

motions grew more accurate, astrology became more ambitious. Claudius Ptolemy, the greatest astronomer in the Greek tradition, wrote the *Tetrabiblos,* a major work which influenced Western astrology. 'Judicial' astrology used the horoscope, or star chart, relating to a person's moment of birth to discover that person's character and the course of their life, or at least their life chances. And the 'aspects' of the heavenly bodies, their relative positions, at the moment of undertaking some enterprise were held to influence its chances of success. Judicial astrology was condemned by two fathers of the Christian Church, St Augustine (354–430) and St Thomas Aquinas (*c.* 1225–74), but taken seriously by many European scientists until the triumph of Newtonian thought in the late seventeenth century.

)))➠ *Newtonian Cosmology, p17*

ZODIAC

An imaginary band around the celestial sphere, lying about 9° to each side of the ecliptic (the apparent annual path of the Sun), in which the Moon and planets (except, sometimes, Pluto) are found. In astrology the name is also given to the constellations found in this band.

)))➠ *Celestial Sphere, p167*

CALENDARS

FOUR CORNERS OF THE WORLD

Accurately determining the length of the year was not easy for early peoples. Most early cultures worked on the basis of a 360-day year, adding extra days after the error became obvious over a series of years. Of all the near-eastern peoples it was the Assyrians and Babylonians, occupying the regions of modern Iran and Iraq, who made the first great innovations in astronomy and geometry. Driven by an official culture which demanded the observation and calculation of astronomical cycles for divinatory purposes, the Babylonians developed the 360° circle (each degree corresponding to a solar day) and exploited its divisibility by 60.

JULIAN CALENDAR

The calendar established by the Roman emperor Julius Caesar in 45 BC continued in use in Christian Europe until 1582. Yet its 365¼ day year was too long by 11 minutes, 14 seconds, and as a result, astronomical events fell earlier and earlier in each year as the centuries passed. This posed a serious problem for Christians, who, like Jews and Muslims, needed reliable calendars to regulate religious worship. The date of Easter, central to Christianity, was calculated in relation to the full Moon on or following the spring equinox. But when was the true equinox: 21 March, as in the calendar, or 11 March, the date on which the astronomical equinox was falling by 1574? A definitive solution was not found until Pope Gregory's 'Gregorian Calendar' of 1582, the calendar used in the West today.

⟫⟫ *Equinox, p168; Solstice, p169*

ISLAMIC CALENDAR

The Islamic calendar is basically a lunar one, which means that observation of the new Moon – at its thinnest crescent – is a fundamental event for time reckoning. However, neither the lunar nor the solar cycles form neatly divisible numbers and from the very start, Muslims, like other peoples, faced the problem of reconciling the two. Although the prophet Mohammed ruled against intercalation on theological grounds, astronomers could not escape the problem; a lunar year of 12 months is roughly 11 days shorter than the solar year.

INTERCALATION

Intercalation is the insertion of extra days or months into the calendar in order to harmonize it with the solar year. In the Gregorian calendar, which is used worldwide today, the extra day added every leap year is an intercalated day.

COSMOLOGICAL THEORIES

FOUR CORNERS OF THE WORLD

All the ancient Near-eastern cultures saw the cosmos as being built in tiers, or levels. The sky was flat and supported in some remote place by pillars at the 'four corners of the Earth'. Above the sky were the 'waters above the firmament', which broke forth to cause the Great Flood of Noah. Then there were the 'waters beneath the Earth', as well as the realms of the dead. This cosmology was shared, in its basic principles, by the ancient Egyptians, Babylonians, Sumerians and Jews. Astronomical bodies were seen as passing beneath the flat Earth at night.

HELIOCENTRIC SYSTEM

Any theory in which the planets, including the Earth, are held to revolve around the Sun. Aristarchus of Samos (310–230 BC) formed a heliocentric theory, but the geocentric (Earth-centred) theory of Claudius Ptolemy dominated astronomy from the second to the sixteenth century. Nicolaus Copernicus (1473–1543) proposed a heliocentric theory in 1543 to explain the variations in the apparent motions of the planets, Moon and Sun. Though it was favoured by increasing numbers of scientists, its final victory depended on new discoveries by Johannes Kepler, Galileo Galilei and Isaac Newton.

)))) *Tychonic System, below; Discovery of Gravity, p17; Kepler's Laws of Planetary Motion, p17*

BELOW: Ptolemy's celestial system had the Earth at its centre, with the Sun, Moon and planets moving around it in perfect circular orbits.

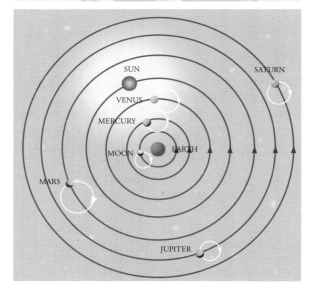

PTOLEMAIC SYSTEM

The peak of Greek astronomical achievement is represented in the encyclopedic *Almagest* of Ptolemy (fl. AD 127–141). Using the works of earlier Greek and Alexandrian astronomers as his basis, Claudius Ptolemy (*c.* AD 120–180) produced a definitive catalogue of 1,022 stars that fixed 48 constellations that we still use. His complex system of circular planetary motions known as the Ptolemaic System reconciled several centuries of recorded observations with the uniform circular motions required by Greek philosophers. Ptolemy firmly established a geocentric (Earth-centred) astronomy for the next 1,400 years.

)))) *Epicycle, below*

EPICYCLE

In early theories of the motions of the solar system, a small circle in which a celestial body moves, whose centre in turn revolves around the Earth by travelling along the perimeter of a larger circle called the deferent. First introduced by Ptolemy, epicycles persisted into the heliocentric (Sun-centred) theory of Nicolaus Copernicus (1473–1543), but were abandoned by Johannes Kepler (1571–1630).

)))) *Equant, below; Kepler's Laws of Planetary Motion, p17*

EQUANT

One of two points equally spaced from the Earth in Ptolemy's geocentric (Earth-centred) model of the solar system. Viewed from the equant, the centre of the planet's epicycle would appear to move at a constant angular rate. The other point was the 'eccentric', about which the centre of the planet's epicycle revolved.

DE REVOLUTIONIBUS

Abbreviated title of the epoch-making book *De Revolutionibus Orbium Coelestium* (*On the Revolutions of the Celestial Sphere*s), published by a Polish churchman, Nicolaus Copernicus (1473–1543). It argued that a heliocentric (Sun-centred) system explained planetary motion more simply than the prevailing geocentric (Earth-centred) one. Nevertheless, like the ancients, Copernicus employed circular motions and a system of epicycles to reproduce the observed motions of the planets, Sun and Moon. He delayed publication of *De Revolutionibus* until the year of his death, fearing academic ridicule. The suggestion that the Earth was hurtling through space affronted common sense and the physics of Aristotle, which dominated the universities.

)))) *Heliocentric System, p15; Ptolemaic System, left; Epicycle, above; Tychonic system, below*

TYCHONIC SYSTEM

Tycho Brahe (1546–1601) was impressed by the mathematical elegance of Copernicus's theory (1543) that the Earth was a planet that moved with the other planets around the Sun, but he could not believe in the physical reality of the Earth's motion. In 1583 he developed a system in which the planets (which did not include the Earth) revolved around the Sun, which revolved around the Earth. The new physics of Isaac

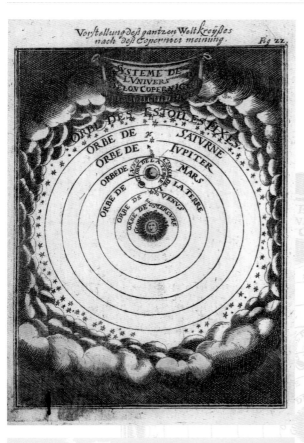

ABOVE: Copernicus's seminal work De Revolutionibus presented his heliocentric solar system, which correctly envisaged Earth and all the other planets revolving around the Sun.

Newton (1642–1727) ruled out the Tychonic system.
>>> *Tycho Brahe, p26; Isaac Newton, p37*

KEPLER'S LAWS OF PLANETARY MOTION

Johannes Kepler (1571–1630) formulated three laws of planetary motion based on the meticulous observations made by Tycho Brahe (1546–1601). The first two were published in 1609 and the third in 1619.

1. The orbit of a planet around the Sun is an ellipse with the Sun at one focus. (Mathematically, every ellipse is defined in terms of two points called the foci, at equal distances from the centre; the greater the separation between the foci, the more elongated the ellipse.) Until that time the planets had been assumed to move in circles.

2. The line connecting a planet to the Sun sweeps out equal areas in equal times. A planet will move much faster when it is closer to the Sun and slower when it is farther away.

3. The square of a planet's orbital period (it's 'year' or the time it takes to go round the Sun) is proportional to the cube of the semi-major axis (half the greatest diameter) of the ellipse.

DISCOVERY OF GRAVITY

In 1666 the young Isaac Newton gained his first insight into gravitation. According to his own account, it was triggered by his realization that the force that pulls an apple from a tree is the same as that which makes the Moon revolve around the Earth. Around 1684 Edmond Halley (1656–1742) visited Newton in Cambridge, and discovered that he had solved the problem of why the Moon moves around the Earth in an elliptical orbit. Halley urged the secretive Newton to publish his findings, and in 1687 *Philosophae Naturalis Principia Mathematica* (Mathematical Principles of Natural Philosophy) appeared, setting forth Newton's revolutionary ideas not only on gravitation but also on mechanics.
>>> *Newtonian Cosmology, p18*

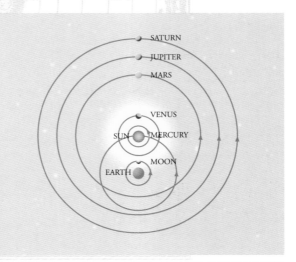

ABOVE: The Tychonic celestial system was an ingenious compromise in which the Earth was central and yet all other planets orbited the Sun.

PHILOSOPHIÆ
NATURALIS
PRINCIPIA
MATHEMATICA.

Autore JS. NEWTON, Trin. Coll. Cantab. Soc. Matheseos
Professore Lucasiano, & Societatis Regalis Sodali.

IMPRIMATUR·
S. PEPYS, Reg. Soc. PRÆSES.
Julii 5. 1686.

LONDINI,
Jussu Societatis Regiæ ac Typis Josephi Streater. Prostant Vena-
les apud Sam. Smith ad insignia Principis Walliæ in Cœmiterio
D. Pauli, aliosq; nonnullos Bibliopolas. Anno MDCLXXXVII.

time seemed to comprise the whole Universe, had the shape of a millstone, with the Sun near the centre (it is in fact about two thirds of the way out). It did not collapse because it was rotating. In the 1920s it was discovered that the Milky Way is just one of billions of galaxies, all rushing apart from each other. By this time Einstein's theories of relativity had supplanted Newton's physics and made modern cosmology possible.

➤➤ *Big Bang, 44–45; General Theory of Relativity, p19*

PRINCIPIA MATHEMATICA

Abbreviated title of *Philosophiae Naturalis Principia Mathematica* (*Mathematical Principles of Natural Philosophy*), epoch-making work of Isaac Newton, published in 1687. Edmond Halley (1656–1742) had urged the secretive Newton to publish, and paid for publication himself. One of the keys to Newton's achievement was his invention of 'fluxions', an early form of calculus. In the work, Newton dismissed the idea of saying what gravity was, and concentrated on elucidating the laws it followed.

SPECIAL THEORY OF RELATIVITY

Physical theory proposed by the German-born physicist Albert Einstein (1879–1955) in 1905 to solve problems in the existing theories of mechanics and electromagnetism. It uses the principle that the laws of physics – in particular, the speed of light in free space – are the same for any two observers that are moving at constant velocity (or are at rest) in relation to each other.

Relativity diverges sharply from Newtonian mechanics at speeds comparable with the speed of light. For example, a spacecraft moving close to the speed of light is reduced in length in relation to observers at rest on the ground. But to an observer in the spacecraft, lengths

NEWTONIAN COSMOLOGY

Isaac Newton (1642–1727) attempted to devise a physical model of the Universe on the basis of his theories of mechanics and gravitation. He recognized that gravity would dominate the Universe on the largest scale, but this led to a problem. If the Universe were finite, everything in it would 'fall down into the middle of the whole space and there compose one great spherical mass'. The fact that this had not happened and apparently was not happening led Newton to the view that the Universe was infinite. However, in an infinite Universe even the slightest unevenness in the distribution of stars would also quickly lead to collapse.

From telescopic observation William Herschel (1738–1822) found that the Milky Way, which at that

ABOVE: Newton's Principia *established the new science of classical mechanics. It covered the laws of motion, the theory of gravity and explained the tides.*

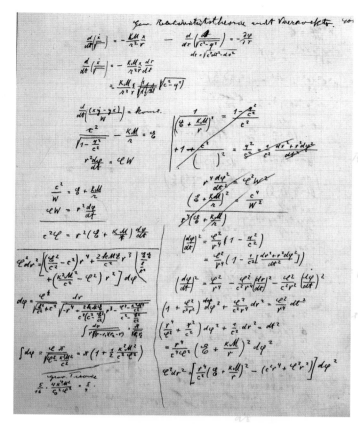

the single dimension of time. Gravity, said Einstein, should be considered as a curvature of space-time caused by bodies with mass. Free bodies move in response to the local curvature of space. The theory can be summed up as: matter tells space-time how to curve, space-time tells matter how to move.

GR made the following predictions, among others, that were subsequently confirmed:
• In a gravitational field, time runs more slowly than in free space. For example, the frequency of light from a star is decreased.
• Light is deflected when passing near a massive object like the Sun.

)))➤ *Special Theory of Relativity, p18; Space-Time, p48; Gravitational Lens, p145*

HUBBLE'S CONSTANT

By 1925, the American astronomer Edwin Hubble had shown the Andromeda 'nebula' to be a galaxy beyond our own, though his estimates for its size and distance were much smaller than those accepted today. With his colleague, Milton Humason, Hubble began to collect data on the distances and motions of similar objects. In 1929, Hubble presented evidence indicating that the farther away such galaxies are, the faster they are receding from us. This was the observational germ of the later Big Bang theory. During the 1930s, Hubble and Humason collected data from increasing numbers of galaxies, and confirmed the correlation. The ratio of speed to distance is called Hubble's constant, and recent measurements indicate that its value is likely to lie in the range 18–24 km/s (11–15 mi/sec) per million light years. Conventionally, the constant is expressed in units of kilometres per second per megaparsec (km/s/Mpc), where 1 megaparsec (Mpc) = 3,260,000 light years. The Hubbble Space Telescope Key Projects team has obtained a value of 72 km/s/Mpc (22 km/s permillion light years).

and distances on the ground are likewise reduced.

Processes on the spacecraft also run slow in relation to those on the Earth, and vice versa. This 'relativity' of space and time accounts for the fact that different observers, even though moving at different speeds, find the same value for the speed of light in free space.

)))➤ *General theory of Relativity, p19; Space-Time, p48*

GENERAL THEORY OF RELATIVITY

Physical theory proposed by the German-born physicist Albert Einstein (1879–1955) in 1915, building on his special theory of relativity of 1905. The general theory of relativity (GR) supplanted Newton's theory of gravitation.

Newton's first law is that free bodies move in straight lines. GR asserts that free bodies follow the equivalent of a straight line not in space but in space-time, a four-dimensional combination of three-dimensional space and

ABOVE: A manuscript showing Einstein's early search for his general theory of relativity, which revolutionized the way scientists thought about space and time.

ASTRONOMICAL DISCOVERIES

HALLEY'S COMET

Before the seventeenth century many astronomers thought that comets were atmospheric phenomena. But Sir Isaac Newton's *Principia, Book III* (1687), showed that comets were astronomical bodies moving in mathematically definable orbits. In 1705, the English astronomer Edmond Halley (1656–1742) published an analysis of cometary records, and argued that the bright comets of 1682, 1607 and 1531 were apparitions of the same body, moving in an elliptical orbit around the Sun. Halley predicted that it would return in 1758 or 1759. A German amateur astronomer, Georg Palitzch, caught the first glimpse of the returning comet on Christmas night, 1758.

LEFT: Edmond Halley, whose observations of the changing positions of stars forced astronomers to reassess their belief in the fixed nature of the Universe.

RADIO ASTRONOMY

In 1932 an American radio engineer, Karl Jansky (1905–49), accidentally detected radio emissions from the Sagittarius region of the Galaxy. In 1937, his compatriot Grote Reber (b. 1911) built a 9-m (31-ft) antenna to detect astronomical radio sources. It was the enormous advance in electronic engineering during the Second World War, however, that really made radio astronomy possible. Sir Martin Ryle (1918–84) at Cambridge pioneered radio telescopes consisting of large fixed arrays of aerials, while Sir Bernard Lovell (b. 1913) developed the steerable dish at Jodrell Bank, Manchester. Radio interferometry (the linking of widely separated telescopes) across the world made it possible to identify the exact positions of radio sources, to match them up with visible light sources and to detect pulsars and other previously unknown objects. Major sources of radio waves include the Sun, interstellar gas clouds, nebulae, supernova remnants and radio galaxies.

ABOVE: The 76-m Lovell Telescope at the Jodrell Bank Observatory.

ASTRONOMICAL LANDMARKS

GREAT PYRAMID AT GIZA

Good surveys of the Great Pyramid of Giza (built *c.* 2700 BC), revealing its accurate astronomical alignment, date from AD 1638. They showed that the pyramid was accurately aligned north–south. It also contains finely cut shafts that point to spiritually significant star positions. The purpose behind these alignments, however, is impossible to know. Nor do we know how many alignments discovered by computer analyses from modern surveys of this and other ancient buildings were ever intended by the original builders. Far-fetched theories about these ancient structures and their builders abound.

BELOW: The monoliths at Stonehenge in the UK are arranged along an axis that is aligned on the point of sunrise at the summer solstice.

STONEHENGE

Ritual monument on Salisbury Plain in southern England, dominated by concentric rings of huge standing stones. The outermost is 33 m (108 ft) wide. Stones were first raised on the site about 2200 BC, having been brought from south Wales. A processional roadway called the Avenue, nearly 3 km (2 miles) long, leaves the site to the north-east, the direction of the summer solstice. Other points of astronomical significance may have been indicated by the alignment of stones.

KEY PEOPLE

ASTRONAUTS

JOHN GLENN (b. 1921)

American astronaut and the first American to orbit Earth. On 20 February 1962, Glenn boarded his Mercury capsule Friendship 7. Minor problems delayed the launch several times, but the countdown was completed at 9.47 a.m. Near the end of his first orbit, Friendship 7 was drifting to the right and Glenn switched to manual to correct it. On the third orbit, mission control noticed a signal suggesting that the heat shield and landing bag were loose. Glenn was told to keep the retrorocket package on after firing so that its straps would hold the heat shield in place. By the time the pack burned away, aerodynamic pressure would keep the shield from slipping. During re-entry Glenn observed large chunks of the retrorocket package ablating. Friendship 7 splashed down safely 22 minutes later. After a career in politics, Glenn made his second and last spaceflight in 1995 when he joined the crew of Space Shuttle mission STS-95. At the age of 77, this made him the oldest person to go into space.

ALAN SHEPARD (1923–98)

American astronaut Alan Shepard was selected as America's first man in space. He named his spacecraft Freedom 7. The launch, from Cape Canaveral in Florida, was originally scheduled for 2 May 1961 but was postponed until 5 May because of bad weather. At 5.15 a.m. EST, Shepard entered Freedom 7 and over four hours later the Redstone rocket propelled him to 8,214 km/h (5,103 mi/h) before shutting down on schedule 142 seconds after lift-off and separating from the capsule. Shepard manoeuvred the craft about all three axes and tested the reaction control systems. Freedom 7 reached a peak altitude of 187 km (117 miles), before Shepard prepared for re-entry, splashing down in the Atlantic Ocean 15 minutes and 22 seconds after launch. Subsequently Shepard commanded the Apollo 14

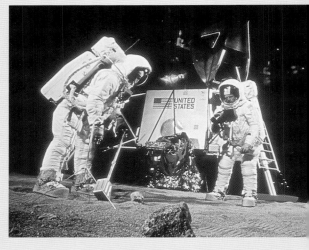

mission which touched down in the Fra Mauro region of the Moon on 5 February 1971.

NEIL ARMSTRONG (b. 1930)

American astronaut. The first man to walk on the surface of the Moon, Armstrong began flying at the age of 14. He had a pilot's licence by the age of 16, before he could drive, and colleagues commented that his love of flight bordered on religious devotion. Armstrong apparently even carried a piece of the original Wright Brothers' flier on board his Gemini 8 mission and treated it like a religious relic. Armstrong joined the Navy as a fighter pilot in 1949, flying 78 combat missions in Korea. On one mission he had to bail out when a wire stretched across a valley took the wing off his F9F-2 jet plane. Back in the US he studied for an aeronautical engineering degree and then joined Edwards Air Force Base as a test pilot in 1955, eventually flying the X-15 rocket plane to a height of 63 km (39 miles) at a speed of Mach 5.74. NASA selected him in 1962 and he flew on Gemini 8 and Apollo 11. Armstrong stayed with NASA after his pioneering Moon walk, but he never flew into space again and left in 1971 to pursue business and academic interests.
))))▶ *Apollo 11, p178*

ABOVE: Edwin 'Buzz' Aldrin and Neil Armstrong on a training exercise for the lunar landing in Apollo 11 – the mission that finally won the Space Race for the USA.

EDWIN 'BUZZ' ALDRIN (b.1930)

American astronaut. The second man on the moon after Neil Armstrong on 20 July 1969 on Apollo 11. Aldrin followed 15 minutes after Armstrong and the pair busied themselves setting up seismic, solar wind and laser-ranging experiments, collecting more samples and taking pictures. They conducted a brief telephone conversation with President Nixon and unveiled a plaque on one of the legs of the Lunar Module (LM) celebrating the achievement of Apollo 'for all mankind'.

JOHN YOUNG (b. 1930)

American astronaut. One of nine astronauts chosen by NASA in October 1962, Young became the first of the group (which included Charles Conrad, Frank Borman and Neil Armstrong) to be assigned to a spaceflight, when he was named pilot of Gemini 3 in April 1964. He was assigned to flight crews almost continuously for the next nine years: in addition to flights on Gemini 3, Gemini 10, Apollo 10 and Apollo 16, he was backup pilot for Gemini 6, backup command module pilot for Apollo 7, and backup commander for Apollo 13 and Apollo 17. He commanded the first space shuttle mission, STS-1, in April 1981 and STS-9 in 1983. He retired from NASA in 2004.

>>> *Gemini Space Program, p175; Apollo Space Program, p176*

YURI GAGARIN (1934–68)

Russian astronaut. The first man to fly in space, orbiting the Earth in his Vostok 1 capsule on 12 April 1961, just a few weeks before the first proposed American manned space flight. The 27-year-old Russian pilot climbed into his Vostok 1 spacecraft and at 04:10 UT was blasted into orbit by Korolev's SL3 rocket. Seventy-eight minutes into the flight, mission control fired the retrorockets and Gagarin began his descent. The Vostoks had no retrorockets for a soft-landing, and so 8,000 m (26,250 ft) above the ground Gagarin ejected from the capsule and parachuted to Earth, landing on the banks of the Volga not far from Engels. Gagarin was killed in a plane crash in 1968.

>>> *Vostok 1, p183*

EUGENE CERNAN (b. 1934)

American astronaut. The second American to walk in space, on flight Gemini 9. During a two-hour space walk, wrestling the umbilical cord and working against an over-stiff pressure suit, Cernan became dangerously overheated and tripled his heart rate to 180 beats per minute. He survived, but his mission had carried out contingency plans for re-entry attempts with a dead crew member abandoned in space. Cernan was also the last man on the moon, on Apollo 17, 1972.

ALEXEI LEONOV (b. 1934)

Russian astronaut. The first man to walk in space when he floated outside the Voskhod 2 spacecraft for 10 minutes on 18 March 1965. After attending an air-force school Leonov served as a fighter pilot before being selected as a cosmonaut in March 1960. Initially a candidate for the first Vostok spaceflight, Leonov was discounted in favour of Gagarin and Titov, who were several inches shorter than him. After his Voskhod 2 flight, Leonov went on to train for the Russian manned circumlunar and lunar landing missions, and could have become the first man to walk on the Moon. The rest of his cosmonaut career was spent on Earth orbital missions, but he did not fly again until the Apollo–Soyuz test flight in 1975. Leonov served as deputy director of the Gagarin Center for cosmonaut training until his retirement in October 1991.

>>> *Apollo-Soyuz Mission, p180*

VALENTINA TERESHKOVA (b. 1937)

Russian astronaut. The first woman in space, on 16 June 1963, the 26-year-old was launched into orbit aboard Vostok 6 and in the next three days circled the Earth 48 times, more than the six American Mercury astronauts combined. In September 1961, inspired by the Vostok flights, Tereshkova had written a letter to the space centre asking to join the cosmonaut team. With her application helped by a background in parachuting, she was selected in March 1962 and, along with four other women, reported to the training centre. Soviet Premier Krushchev personally picked her out for the Vostok 6 flight because of her working-class background. Famed as a heroine of the women's movement in Soviet society, she went on to a career in politics and married fellow cosmonaut Andrian Nikoleyev. Tereshkova later earned a Candidate of Technical Sciences degree (in 1976) and was eventually promoted to the rank of major general, retiring in March 1997.

ASTRONOMERS

EUDOXUS (407–355 BC)

Greek astronomer. Eudoxus tried to explain the observed motions of the planets by devising a geometric model which contained 27 spheres, representing the motion of the planets in relation to the zodiac, while rotating around the Earth. The path taken by a planet against what was then thought to be a fixed background of stars could be observed in this model as a figure-of-eight curve, which Eudoxus termed a *Hippopede* (Greek: 'Horsefetter'). Eudoxus' system was the ancestor of the epicycle and eccentric circles system later developed by Hipparchus (fl. 160–125 BC).

ERATOSTHENES (c. 276–195 BC)

Greek astronomer. Eratosthenes pioneered a technique which measured the size of the Earth. He knew that at midsummer the Sun shone right to the bottom of a deep well in the town of Syene, near the modern Aswan, and hence was directly overhead. Yet in Alexandria, just over 7° to the north, the Sun cast shadows. Eratosthenes also knew that as the rays of the Sun striking the Earth were parallel, the shadow angle cast in Alexandria must be the same as that produced by radial lines drawn from Syene and Alexandria to the centre of the Earth. As Syene and Alexandria were 5,000 Greek stadia apart, he deduced that they were one-fiftieth of a circle apart, so the circumference of the Earth must equal 250,000 stadia. Although there is some debate about the exact length of Eratosthenes' stadia, his figure produces an Earth circumference of about 47,000 km (29,000 miles). The modern value is 40,074 km (24,902 miles).

HIPPARCHUS (190–120 BC)

Greek astronomer. One of his most remarkable achievements was the construction of the Hipparchian Diagram, from which he attempted to

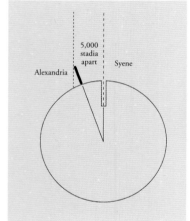

5,000 stadia apart
Syene
Alexandria

calculate the distances and respective sizes of the Moon and the Sun, from the ways in which the fast-moving (and hence closer) Moon exactly covers the slower-moving Sun during a total eclipse. Hipparchus calculated that the Moon's maximum distance was 67.5 earth radii, which is astonishingly close to the correct value. He is said to have discovered the precession of the equinoxes, and constructed the first star catalogue. As an observer, a theoretician, and a utilizer of Babylonian science, Hipparchus consolidated and redirected Greek astronomy.

CLAUDIUS PTOLEMY (c. AD 120–180)

Greek astronomer. Using Hipparchus's earlier works as his basis, Ptolemy produced *Magna Syntaxis* ('Great Syntax') or, in Arabic, the *Almagest of Ptolemy* and his celestial system, known as the Ptolemaic system. His Earth-centred cosmology harmonized with the physics of Aristotle, provided quantities from which calendars could be calculated, and was used equally by Christian, Jewish, Arabic and Hindu astronomers.

⟫ *Ptolemaic System, p16*

RICHARD OF WALLINGFORD (c. 1292–1336)

English astronomer. One of the first clock designers, Wallingford, Abbot of St Albans, built an extraordinary mechanism which not only told the time, but which also rotated circles to replicate the motions of the Ptolemaic Universe. To reproduce the complex motion of the Moon he used elliptical and differential gears some three centuries before Jeremiah

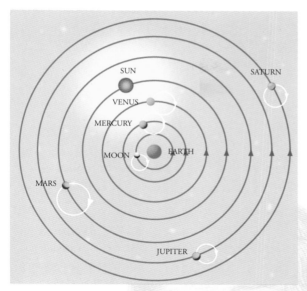

ABOVE: Ptolemy's celestial system had the Earth at its centre, with the Sun, Moon and planets moving around it in perfect circular orbits.

Horrocks (1619–41) actually proved the lunar orbit to be elliptical. Astronomers and inventors across Europe began to build clocks, to tell the time and to replicate the celestial motions. In 1364, Giovanni de Dondi built an even more complex astronomical clock. Detailed descriptions of the Wallingford and Dondi clocks still survive.

ULUGH BEG (1394–1449)

Arabic astronomer. In 1420 he built a great three-storey, drum-shaped observatory, 30 m (100 ft) high, housing a 40-m (130-ft) radius masonry sextant set in the plane of the meridian. Alongside the graduated edge of this sextant, each brass degree spanned 68 cm (27 in). Astronomers observed the Sun, Moon or stars through a small hole at the geometrical centre of the instrument by attaching a finely graduated sliding pinhole eyepiece and scale between each degree. With this colossal sextent Beg determined the length of the solar year to within a minute of time. He also composed a catalogue which gave the precise positions of 1,018 stars.

GEOFFREY CHAUCER (c. 1343–1400)

English poet. Although his fame rests on his poetry, Chaucer was the first man to write a scientific book in the English language, rather than in Latin. His *Treatise on the Astrolabe* (c. 1391) is a detailed manual for the use of that Arabic astronomical instrument, already commonplace across Europe, and which coordinated astronomy, angle measuring, and time finding. The brass astrolabe was used by calendar calculators, university teachers and students studying astronomy, in the same way that a modern person might use a planisphere or a computer. Chaucer is also believed to be the author of the *Equatorie of the Planets* (c. 1392), which describes a brass instrument, the equatorium, used for planetary calculation.

)))⫸ *Astrolabe, p12*

COPERNICUS, NICOLAUS (1473–1543)

Polish astronomer. Nikolaj Koppernigk, latinized to Copernicus, was born into a prominent Polish family in 1473. Nicolaus's education was supervised by his powerful uncle, Lucas Waczenrode, Bishop of Ermeland, who no doubt recognized his nephew's abilities and saw their future use to Poland. Nicolaus was first sent to the University of Cracow in 1491, where he received a Latin training in classical literature, law and theology, and then, in 1496, to Italy. He studied in Italy until 1503, gaining doctorates in law and medicine, and acquiring skills that would be important to his perceived administrative career back in Poland. He also became fascinated

ABOVE: Nicolaus Copernicus was the first to suggest a heliocentric model of the Universe, as opposed to the long-established and accepted geocentric model.

with the new Greek learning of the Renaissance humanists, acquiring the familiarity with Greek astronomy and its problems that would one day make him famous. Through his uncle's influence, Copernicus obtained the canonry of Frauenberg Cathedral, which gave him a comfortable income for life. Here, amidst his public and ecclesiastical duties, Copernicus was to spend the next 40 years of his life. And it was at Frauenberg Cathedral that he would study the pre-Ptolemaic Greek astronomers, make some observations, and quietly develop his extremely influential heliocentric theory.

TYCHO BRAHE (1546–1601)

Danish astronomer. One of the leading astronomers of the Renaissance, as a youth he studied law at Leipzig, Wittenberg and several other European universities, but even by the age of 16, a passion for astronomy was uppermost in his mind. On 24 August 1563 Brahe made his first recorded observation of Jupiter and Saturn in conjunction. He also collaborated with Paul Hainzel of Augsburg in the construction of a quadrant for accurately observing the planets. Tycho's eccentric brilliance was recognized by King Frederick II of Denmark, who presented him with the Island of Hven on which he built his famous 'Castle of the Heavens', Uraniborg. Tycho lived and worked at Uraniborg between 1576 and 1597, amidst the research assistants and technicians of Europe's first scientific research academy. His aim was to prove astronomical theory by practice and mathematical analysis. Court politics caused him to leave Denmark, however, and he died in Prague in 1601.

)))⫸ *Tychonic System, p16*

GALILEO GALILEI (1564–1642)

Italian astronomer. Galileo studied at Pisa University and, after abandoning medical studies, took to mathematics, becoming a lecturer at Pisa in 1588. It was in Pisa that Galileo made his first researches into swinging pendulums and falling bodies. In 1591 he became professor of Mathematics at Padua, which was

then the finest scientific university in Europe. It was here, in 1609–10, that he made telescopic discoveries which, when published in his *Siderius Nuncius* ('Starry Messenger'), revolutionized astronomy and made him an international celebrity. His disputed co-discovery and anti-Aristotelian interpretation of sunspots with Christopher Scheiner after 1612 made him many enemies amongst the intellectual Jesuit order, which was probably instrumental in engineering his trial in 1633.

EDMOND HALLEY (1656–1742)

English astronomer. In 1715 Halley published a paper describing six stars that had altered considerably over time. Brahe's supernova of 1572 had disappeared altogether, whereas others, Mira, for example, varied in brightness. In 1716 he drew attention to the existence of six 'lucid spots' or telescopic nebulae. In 1718 he further discovered that three bright stars – Aldebaran, Sirius and Arcturus – had shifted position in relation to other stars since Ptolemy's time. Halley went on to question how stars could be found to have changed in position and brightness, and how they related to the nebulae. And why, if the Universe contained uncountable millions of telescopically visible stars in all directions, did the sky ever go dark at night? This problem would later be redefined as Olbers' Paradox. Best-known for his successful prediction that the comet which now bears his name would return in 1758 or 1759, Halley was instrumental in encouraging Newton to publish his *Principia*.

))))➤ *Isaac Newton, p37; Olbers' Paradox, p47*

JOHANNES KEPLER (1571–1630)

German astronomer. A Lutheran Protestant from Weil in Catholic South Germany, Kepler's talent won him a place to train for the ministry at Tübingen, and while never ordained, his love of astronomy was imbued with a powerful religious conviction. After lecturing at Gratz in Austria (where the idea of the geometrical solids and the planets came to him), Kepler was taken up by Tycho Brahe (1546–1601), one of the leading astronomers of the Renaissance. Following Tycho's death he succeeded Tycho as Imperial Mathematician. He became Mathematics professor at Linz in 1612, and at Ulm in 1627 he published his *Rudolphine Tables*, based on Tycho's observations. From his analysis of Tycho's observations, he established three laws of planetary motion. He predicted the transits of Mercury and Venus in 1631. Kepler's optical researches also led him to invent the telescope eyepiece that bears his name.

))))➤ *Kepler's Laws of Planetary Motion, p17*

CHRISTIAAN HUYGENS (1629–95)

Dutch mathematician and astronomer who proposed that light was a form of wave motion. In 1690 Huygens showed that each point on a wavefront, (the surface connecting crests of a wave), could be regarded as a source of 'secondary wavelets' that spread out in all directions. Huygens' principle provides a powerful model for the propagation of light and is still used in teaching the physics of diffraction. Huygens discovered Saturn's largest satellite, Titan, in 1655. The following year, he produced the first correct explanation of Saturn's rings, publishing a full

BELOW: Johannes Kepler's investigations into the orbit of Mars revolutionized planetary theory.

account of his findings in Systema Saturnium in 1659. He was also the first astronomer to sketch genuine features on the surface of Mars and to make a reasonable estimate of the planet's rotation period.

SIR WILLIAM (1738–1822) AND CAROLINE (1750–1848) HERSCHEL

German-born astronomers who were based in England. William's great discovery was the previously unknown planet Uranus in March 1781. His sister Caroline also revealed herself to be an astronomer of great talent. As well as assisting William, she discovered eight comets in her own right, and after his death in 1822 Caroline was honoured by the Royal Astronomical Society, receiving its prestigious Gold Medal, for completing his work.

FRIEDRICH BESSEL (1784–1846)

German astronomer. In 1838 Bessel announced that the star 61 Cygni had a parallax that placed it 10.3 light years away, only one light year closer than the currently accepted distance. As the precision of telescopes improved, so had hopes of measuring the microscopic parallax shifts that stars would show if they were at distances comparable to those Newton had estimated for Sirius, hence Bessel's successful result.

URBAIN LE VERRIER (1811–77)

French astronomer whose analysis of the motion of the planet Uranus led to the discovery of the planet Neptune in 1846. A similar calculation was made by John Couch Adams. Le Verrier's analysis of the motion of Mercury led him to predict, erroneously, the existence of a planet closer to the Sun than Mercury.

ABOVE: Sir William Huggins, who was a pioneer in recording stellar spectra photographically during his 40-year career. By 1880, the new science of astrophysics was transforming cosmology.

Maunder minimum (1645–1715) and the Spörer minimum (1450–1540), confirmed by radiocarbon dating of tree rings, as being times of low solar activity. The Maunder and Spörer minima also coincide with prolonged cold spells, hinting at the Sun's influence on Earth's climate.

)))➤ *Maunder Minimum, p71; Sunspots, p70*

JACOBUS KAPTEYN (1851–1922)

Dutch astronomer who in 1906 set out to map the structure of the Galaxy. Collaborators counted the numbers of stars of different brightness in some 200 selected areas of sky and measured their motions. Kapteyn wrongly concluded that the Sun lay close to the centre of the Galaxy. However, Harlow Shapley (1885–1972), a US astronomer, later demonstrated around 1918 that the Sun lay well away from the centre

)))➤ *Harold Shapley, p29*

JOHN DREYER (1852–1926)

Danish-born astronomer. Dreyer worked as an assistant at the observatory of Lord Rosse in Ireland, where he observed faint nebulae and star clusters with Rosse's huge telescope, and then at the Armagh Observatory, where he was director from 1882 to 1916. At Armagh, Dreyer continued his interest in what are now termed deep-sky objects, publishing in 1888 the *New General Catalogue of Nebulae and Clusters of Stars* (popularly known as the NGC), a revision and expansion of a catalogue published earlier by John Herschel which contained 7,840 objects. In 1895 and 1908 Dreyer published two supplements, called the *Index Catalogues*, which added over 5,000 newly discovered objects.

SIR WILLIAM HUGGINS (1824–1910)

English astronomer. Huggins showed in 1863 that stars are made of the same elements as the Sun and detected gas in six nebulae in 1864. The difference between stellar and gaseous nebulae became obvious when examining them through a spectroscope. Huggins and his wife Margaret, together with Henry Draper (1837–82) in New York, photographed the spectra of thousands of objects over four decades, and played a major role in establishing the new science of astrophysics.

E.W. MAUNDER (1851–1928)

British astronomer who created the 'Butterfly Diagram', which depicts how the latitude and size of sunspots changes during the sunspot cycle (a phenomenon recorded by the German astronomer Gustav Spörer, 1822–95). This led to comparing different sunspot cycles and to the realization, first noted by Spörer, that historical records indicated long periods when there seemed to be essentially no sunspots at all. We now recognize the

PERCIVAL LOWELL (1855–1916)

American astronomer who founded the Lowell Observatory in Flagstaff, Arizona. He became famous for his observations of the planet Mars. He observed the 'canals' seen by Giovanni Schiaparelli and believed that they were immense artificial structures, built by a native Martian civilization in order to irrigate their vast deserts with water from the polar caps. However these canals were an optical illusion, and few of them corresponded to real features on the Martian surface. Lowell also initiated the search for the planet Pluto.

Although his calculations were spurious, they led the way to Clyde Tombaugh's discovery of the ninth planet in 1930, 14 years after Lowell's death.

GEORGE ELLERY HALE (1868–1938)

American astronomer. Hale developed the first spectrohelioscope for studying the solar surface in the light of calcium and hydrogen. In 1892 he became first director of the University of Chicago's Yerkes Observatory, when he persuaded streetcar magnate Charles Yerkes to pay for what is still the world's largest refracting telescope, with a 1-m (40-in) lens. Hale used this telescope to make observations of the Sun's chromosphere, but atmospheric conditions at Yerkes were not ideal for solar work and Hale eventually relocated to Mount Wilson near Los Angeles. There he built two major solar-tower observatories and began important programs of photographic observation of the Sun. More significant for astronomy in general, he had a flair for gaining funds to build large optical telescopes and his legacy is the 2.5-m (100-in) Mount Wilson and the 5-m (200-in) Palomar telescopes. He was also a founder, in 1895, of the prestigious Astrophysical Journal.

)))**)** *Mount Wilson Observatory, p163*

HENRIETTA LEAVITT (1868–1921)

American astronomer. Between 1908 and 1912, Leavitt, in her analysis of photographic plates of the Small Magellanic Cloud taken by the Harvard Observatory's telescope in Peru, discovered a curious thing: all the variable stars which showed a light-output curve similar to Delta [δ] Cephei – or Cepheid variables – shared a mathematical relationship. The brighter the absolute magnitude of the Cepheid, the longer it took for the particular star to go through its light cycle. Because all the stars in the Magellanic Cloud were virtually the same distance from Earth, they provided an ideal laboratory in which to quantify the period-luminosity patterns of bright and dim Cepheids.

EJNAR HERTZSPRUNG (1873–1967)

Danish astronomer born in Roskilde, Denmark, a late-comer to astronomy at the age of 29. He produced a wealth of valuable results on magnitudes, colours, parallaxes and proper motions of stars, including visual binaries and variables. He was the first to realize that for stars of a particular colour there was a wide range in luminosity – he therefore discovered giants and dwarfs.

)))**)** *Luminosity, p116*

HARLOW SHAPLEY (1885–1972)

American astronomer who demonstrated around 1918 that the Galaxy was far larger than the estimate of Jacobus Kapteyn (1851–1922), a Dutch astronomer, and that the Sun lay well away from the centre of the Galaxy. American Robert Trumpler (1886–1956), showed that Kapteyn's results were badly affected by interstellar dust, which gave the erroneous impression that their numbers fell off equally in all directions around us. Shapley's determination of the size of the Galaxy was based on the distribution of the globular clusters.

)))**)** *Jacobus Kapteyn, p28*

EDWIN POWELL HUBBLE (1889–1953)

American astronomer. Using Cepheid variable stars, Hubble determined the distance to the Andromeda galaxy in 1924. He proposed an evolutionary sequence beginning with near-circular elliptical galaxies and moving through increasing ellipticity to reach a basic form of spiral galaxy. The sequence, he believed, continued as two parallel forms of spiral, forming the famous tuning fork diagram. Although the evolutionary part of this classification is no longer believed to be correct, Hubble's basic forms are still used to define galactic types. In 1929 Hubble proposed what is now known as Hubble's law, implying that we live in an expanding Universe.

)))**)** *Tuning Fork Diagram, p150*

Percival Lowell, known for his theories about mythical Martian 'canals'.

ABOVE: Sir Bernard Lovell, one of the pioneering fathers of radio astronomy, in front of the Jodrell Bank radio dish he campaigned for in the 1950s.

WALTER BAADE (1893–1960)

German-born US astronomer responsible for the idea of stellar populations, which led to the solution of a major riddle regarding the age of the Universe. Before the 1950s, geologists argued (correctly) that Earth was about 4.5 billion years old, whereas astronomers argued (incorrectly) that the age of the Universe was about 2 billion years. The astronomers' argument was based on the distances to other galaxies, from which the Hubble constant and age of the Universe could be calculated. Baade realized that there were two classes of pulsating variable star on which the distance scale was based, each of which had a different relationship between their period of pulsation and luminosity. Use of the incorrect relationship had led astronomers to underestimate the distances of galaxies. Baade's revision of the astronomical distance scale doubled the known size (and hence age) of the Universe, and subsequent observations have increased these figures still further.

CECILIA PAYNE-GAPOSCHKIN (1900–79)

British-born astronomer. Author of a thesis, *Stellar Atmospheres*, in which she used the calculations of the Indian astrophysicist Meghnad Saha (1893–1956) to interpret the spectra of stars according to temperature and to deduce their chemical compositions. She later demonstrated that hydrogen was the main constituent of stars, which had not been appreciated until then.

JAN HENDRIK OORT (1900–92)

Dutch astronomer who made many significant contributions to galactic astronomy. By analyzing the motions of distant stars, he was able to quantify the rotation of the Galaxy and determine the Sun's distance from its centre. He discovered there is a spectral line at radio frequencies that we can detect, and his student Hendrik van de Hulst (1918–2000) calculated that it is emitted by atomic hydrogen at a wavelength of 21 cm (8 in). This is valuable because atomic hydrogen is common in the disks of galaxies and because absorption is negligible at radio wavelengths. In the 1950s Oort and collaborators used 21-cm (8-in) hydrogen observations to make the first map of the spiral arms of our Galaxy. Oort is also famous for his proposal in 1950 that there is a cloud of comets around our Solar System, now termed the Oort Cloud.

ERIK HOLMBERG (1908–2000)

Swedish astronomer. Holmberg developed techniques for measuring basic properties of galaxies such as their brightnesses and sizes. In 1941, before the invention of the computer, he undertook the first N-body simulation. Since the strength of a gravitational force decreases with distance in the same way as the apparent brightness of a light source, Holmberg used 74 light bulbs to represent a scale model of two galaxies. The amount of light arriving at the location of each bulb represented the gravitational pull on that part of the system. Holmberg was thus able to simulate the motions of the galaxies and demonstrate that colliding systems produce tidal tails and eventually merge.

SIR BERNARD LOVELL (b. 1913)

English astronomer who conducted radar research during the Second World War. In December 1945 he installed ex-military radar equipment at the University of Manchester's botanical research station at Jodrell Bank in

Cheshire, and studied radar echoes from daytime meteor showers, and cosmic rays from the Milky Way and beyond.

Lovell's project to construct a vast 76-m (250-ft) steerable dish gained momentum with the imminent birth of the Space Age and was rushed to completion by 1957 – just in time to track Sputnik 1's carrier rocket. The Russians were delighted when Lovell used the telescope to confirm that their Luna 2 probe had reached the Moon, and NASA used it to contact their first deep-space probe, Pioneer 5. Lovell directed the observatory until 1981 and the telescope has since been named after him. Refitted in the 1990s with a new reflective surface, it is still among the most powerful radio telescopes in the world.

SIR MARTIN RYLE (1918–84)

British astronomer. A pioneer of radio astronomy, in the 1950s, Ryle discovered how to use several small radio telescopes to imitate the performance of a larger instrument, a technique he called aperture synthesis. He built two major aperture synthesis telescopes at Cambridge, the One-Mile Telescope and the Five-Kilometre Telescope (which is now known as the Ryle Telescope).

He was a leading proponent of the Big Bang theory for the origin of the Universe, based upon his own surveys of distant radio sources which indicated that radio galaxies had been more powerful in the past and that the Universe must have been evolving.

ALLAN SANDAGE (b. 1926)

American astronomer who was a major figure in observational astronomy. Sandage's early work was a continuation of Hubble's attempt to measure the geometry of the Universe using optical observations of distant galaxies. In 1960 he made the first optical identification of a quasar. With Thomas Matthews he found a faint optical object coincident with the compact radio source 3C48. Sandage noted the object's unusual spectrum, which was soon shown to be due to a large redshift. He continued to engage in a vigorous debate over the value of Hubble's constant and hence the age of the Universe, arguing for a low value of the constant and a large age for the Universe.
Hubble's Constant, p19

RICCARDO GIACCONI (b. 1931)

Italian astronomer who discovered the first cosmic X-ray source and built two X-ray satellites. In 1959 Giacconi joined American Science and Engineering, a small research company which discovered the first source of cosmic X-rays in 1962. His group went on to build the Uhuru satellite (1970) and in 1973 he played a key role in developing the Einstein X-ray observatory. In 2002, he won the Nobel Prize in Physics for his pioneering contributions to astrophysics.

JOCELYN BELL (b. 1943)

English astronomer. As a student working on her doctoral thesis at the Mullard Radio Astronomy Observatory, Cambridge, England in 1967, Bell studied interstellar scintillation, the 'twinkling' of radio waves from distant sources by intervening interstellar material, as a method of searching for quasars. Bell noticed some strange squiggles on the chart depicting readings from a particular part of the sky, which, when examined in more detail, turned out to be pulses of radio energy repeating precisely every 1.3373 seconds. Soon other sources were found. Bell and her supervisor, Professor Antony Hewish (b. 1924), announced the discovery of these sources, called pulsars, in 1968. Other astronomers soon established that they were rapidly rotating neutron stars.

BELOW: Allan Sandage, Edwin Hubble's former assistant and now one of the world's leading cosmologists.

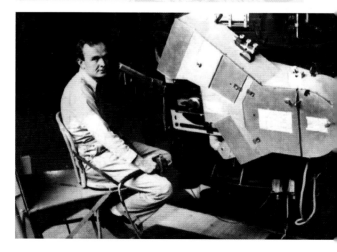

PHYSICISTS

RENÉ DESCARTES (1596–1650)

French physicist. Descartes developed what became known as the Mechanical Philosophy, a comprehensive system of physics which argued that all movement was a product of impacts from corpuscles, or tiny physical bodies. Corpuscles of different size were the basis of everything, and space was filled with the smallest of these. Swirling 'vorticles' in this universe of corpuscles could move the planets and comets or transmit light, in accordance with exact mathematical laws. The drawback with Descartes' physics was its highly speculative character, as 'thought experiments' ultimately predominated over actual experiments.

HENRY CAVENDISH (1731–1810)

English physicist. The aristocratic Cavendish was a skilled experimenter who made the first experimental measurement of the gravitational constant, known as G, in 1798. He achieved this by mounting a small lead sphere at each end of a metal bar suspended at its centre by a wire. He placed two larger lead spheres nearby so that the smaller masses would be attracted towards them. By measuring the tiny angle through which the bar turned, Cavendish was able to calculate the force on the spheres and so work out the value of G. With G known, he could then calculate the mass and density of the Earth for the first time.

JOSEPH VON FRAUNHOFER (1787–1826)

German physicist and instrument-maker recognized as the founder of astronomical

spectroscopy. In 1814 he devised a way of measuring the optical properties of glass by using the bright yellow emission line in the spectrum of sodium as a standard. He devised the modern spectroscope by using a telescope to study the spectrum produced by a prism. He found that the spectrum of the Sun was crossed by numerous dark lines, one of which corresponded in wavelength to the emission line found in the spectrum of sodium. Fraunhofer went on to catalogue more than 500 of these absorption lines – now known as Fraunhofer lines – and the system of letters he used to label the more prominent lines is still used. In 1821 he constructed the first diffraction grating and used it instead of a prism to form a spectrum and make precise measurements of the wavelengths of the solar absorption lines.

WILLIAM THOMSON, LORD KELVIN (1824–1907)

British physicist who is best remembered for the introduction of the absolute scale of temperature which bears his name. His scale is independent of any physical substance, in line with his desire to create an international system of standards. Thomson's early work was in electromagnetism where he gave a mathematical basis to the discoveries of Michael Faraday, paving the way for James Maxwell's great synthesis.

Thomson also helped develop the theory of thermodynamics, and debated the age of Earth, which he estimated from its rate of cooling to be as low as 20 million years – no-one then knew that the temperature of Earth is maintained by radioactive decay.

TOP: Henry Cavendish was the first to measure the gravitational constant.
LEFT: Jospeh von Fraunhofer was the first person to study systematically the dark absorption lines in the Sun's spectrum that now bear his name.

JAMES CLERK MAXWELL (1831–79)

Scottish physicist. Maxwell devised a mathematical representation of the electric and magnetic fields and how they depended on each other, written today as four simple equations called Maxwell's equations. He showed how all the observed phenomena of electricity and magnetism could be understood. By thinking of space as an elastic medium, he discovered that electric and magnetic fields could travel through space in the form of a wave. The speed of the wave could be predicted and came out to be very close to the speed of light. Maxwell proposed in 1864 that light was nothing less than one form of electromagnetic wave, and that an infinite range of 'invisible light' of longer and shorter wavelengths should also exist.

LEFT: James Clerk Maxwell's theory of electromagnetism predicted the existence of electromagnetic waves more than 20 years before their discovery.

HEINRICH HERTZ (1857–94)

German physicist who discovered radio waves and showed that they travelled at the speed of light. In 1888, in order to test Maxwell's theory, Hertz generated electromagnetic waves by causing an oscillating spark to jump between two electrodes at the focus of a parabolic reflector. He detected them with a similar apparatus that produced a small spark in response to the waves. This new type of electromagnetic radiation, now called radio, was the first to be discovered after light, infrared and ultraviolet radiation.

MAX PLANCK (1858–1947)

German physicist. Planck had proposed his radical quantum theory in order to solve the black-body radiation problem, but he half suspected that it was just a mathematical trick rather than a true description of radiation. He spent many years attempting to find a way around his own theory – unfortunately without success.

MACH'S PRINCIPLE

Ernst Mach (1838–1916), an Austrian physicist and philosopher, argued that the laws of physics and the physical constants, such as the speed of light, were determined by the matter in the Universe. Therefore, if the overall properties of the Universe, such as its density, changed then the laws and the constants should also change. In particular he argued that the inertial mass of a body, i.e. its reluctance to be accelerated, is not a property of the body itself, but is an interaction between the body and the rest of the matter in the Universe. For this inertial effect to be true, i.e. so that it is not solely the matter in the Solar System which affects local objects, but distant matter as well, then the force must be very long range.

⏵ *Steady State Theory, p48*

RIGHT: Max Planck, one of the twentieth century's most influential physicists.

VICTOR HESS (1883–1964)

Austrian physicist who discovered cosmic rays in 1912. In 1910 Hess investigated the origin of background radiation that appeared in radioactivity experiments, even within lead-shielded containers. Hess demonstrated in a series of experiments involving hazardous balloon ascents that the background radiation was due to extremely powerful cosmic rays coming from space. This illustrated that the cosmic rays were far stronger than those generated in laboratories and so penetrating that they could be detected through a metre of lead or a 500m (1,640ft) depth of water. Investigation of the nature of cosmic rays led directly to the discovery of the positron by Carl Anderson, and he shared with Hess the Nobel Prize for physics in 1936.

NIELS BOHR (1885–1962)

Danish physicist. The father of quantum mechanics, in 1911 Bohr developed a model of the atom based on electron orbits that obeyed the rules of the then-new quantum theory. He was awarded the Nobel Prize for this work in 1922.

During the 1920s Bohr headed a new institute in Copenhagen which became the centre for work on quantum mechanics. Bohr famously entered into a long-running debate with Einstein over the meaning of the new theory, developing what has become known as the Copenhagen interpretation.

ALBERT EINSTEIN (1879–1955)

Physicist. Einstein was born in Germany, but took first Swiss and then US citizenship. He devised the two theories of relativity and made important contributions to quantum physics. As a boy he questioned what a beam of light would look like if he could catch up with it. In 1905, while working in the Swiss patent office, Einstein published three landmark papers: one in which he introduced the concept of the photon (which won him the Nobel Prize in 1921) and one in which he set out the theory of special relativity. The third was on Brownian motion which contained evidence for the atom. But it was not until 1909 that he secured his first academic post, at Zurich University, and became an established member of the scientific community.

The general theory of relativity, published in 1916, explained gravity not as a force between particles but as a distortion in the fabric of space-time. The latter part of his life was spent unsuccessfully trying to unify gravity with electromagnetism. In 1933 Einstein left Nazi Germany for

BELOW: Niels Bohr, considered by many to be the father of quantum mechanics. He won the Nobel Prize in 1922 for his model of the atom based on electron orbits.

Princeton, where he was to remain. In 1939 he alerted the US government to the possibility of the Nazis developing an atomic bomb and later campaigned for nuclear disarmament.

GEORGE GAMOW (1904–68)

Russian physicist. Trained in nuclear physics in Leningrad (now St Petersburg), Gamow moved to the United States in 1934. He spent several years working on radioactivity and the quantum theory of the atomic nucleus. This led him to study the nuclear reactions that supply the energy for stars.

In the late 1940s, while researching the physics of the Big Bang, Gamow realized that the radiation emitted when the Universe was a hot, uniform gas should still be detectable, but redshifted to a temperature only a few degrees above absolute zero. It took a further two decades for the American physicists Arno Penzias (b. 1933) and Robert Wilson (b. 1936) to discover cosmic background radiation, and then only by accident.

ANDREI SAKHAROV (1921–89)

Russian physicist. In 1967 Sakharov argued that a 'matter Universe' could have emerged from the initial creation process if that process had produced a very small excess of matter over antimatter. Matter and antimatter would later have annihilated, and at the end of the annihilation process the slight excess of matter would have remained to evolve into the Universe we see now.

GARY FLANDRO (b. 1934)

American physicist. Whilst working at the Jet Propulsion Laboratory in California in 1964, Flandro was interested in planetary positions

Two of the eminent theoretical physicists who had previously occupied this chair were Newton (1642–1727) and Paul Dirac (1902–84). Hawking's work has been an attempt to synthesize the interests of gravity and quantum mechanics. After early work on relativity, Hawking concentrated on the problems of gravitational singularities, publishing important work on black holes and the Big Bang. In particular he has shown that black holes are not an irreversible energy sink, but by the processes of quantum mechanics slowly evaporate by the emission of thermal radiation. The rate of evaporation is inversely proportional to the mass of the black hole, so as the black hole diminishes in size the rate of mass loss increases. In 1988 Hawking published *A Brief History of Time*, an outstandingly successful account of the current thinking in cosmology. It was another major achievement for a man afflicted by a crippling motor neurone disease, which has confined him to a wheelchair and requires him to talk by means of a voice synthesizer.

⟫⟫▶ *Big Bang, p44; Hawking Radiation, p51; Black Holes, p54*

LEFT: *Einstein saw that gravity should be understood not as the force between masses but as the effect of distortions in the four-dimensional fabric of space-time.*

and computed a series of theoretical sling-shot missions. He discovered that at the end of the 1970s, a 175-year planetary alignment would enable a spacecraft to visit all the outer planets in one 12-year flight. From these calculations, the 'Grand Tour' of the Solar System emerged, later becoming successful Voyager missions.

STEPHEN HAWKING (b. 1942)

English theoretical physicist. Hawking was born in Oxford (on the 300th anniversary of Galileo's death). Having graduated from Oxford University he moved to Cambridge to study for his PhD. He was made Lucasian Professor of Mathematics at Cambridge in 1980.

RIGHT: *Stephen Hawking, who demonstrated that naked black holes (those without accretion disks) can also emit thermal radiation. The smaller the black hole, the greater the gravitational tidal force it produces. This explains why a smaller black hole is capable of greater radiation than a larger black hole.*

ASTROPHYCISISTS

ARTHUR EDDINGTON (1882–1944)

English astrophysicist. A brilliant student, Cumbria-born Eddington became Plumian Professor at Cambridge in 1913, where he spent the remainder of his career. One of the early and most authoritative proponents of Albert Einstein's (1879–1955) theory of relativity, he was closely involved in one of the expeditions to photograph the solar eclipse of 1919 which demonstrated the bending of light by a gravitational field, in this case the Sun's. Eddington worked in many areas of astronomy, such as stellar dynamics and the nature of the interstellar medium, but one in which he played a major role was in the application of physical and mathematical principles to describe the structure of a star.

This was encapsulated in his benchmark book *The Internal Constitution of the Stars* (1926). Towards the end of his life he devoted his considerable intellectual powers to a search for connections between the fundamental constants of nature.

BELOW (right): Astrophysicist Subrahmanyan Chandrasekhar.

FRITZ ZWICKY (1898–1974)

Swiss astrophysicist who first coined the word supernova in 1934 to describe a class of stellar outbursts different in scale and nature from the 'novae' known at that time. Zwicky suggested a link between supernovae and neutron stars, more than 30 years before the first neutron stars were discovered. In 1933 he was studying the Coma cluster of galaxies and found that the speeds of the individual galaxies were so high that they should have escaped long ago. He concluded that the amount of matter in the cluster was much higher than could be accounted for by the visible galaxies. Sometimes called 'missing mass' this matter is now usually called 'dark matter.'

SUBRAHMANYAN CHANDRASEKHAR (1910–95)

Indian theoretical astrophysicist. Chandrasekhar graduated in Madras in 1928 and went to Cambridge in 1930 to work on the application to stellar structure of a new theory of gases. He showed that the matter in the interiors of white dwarfs was degenerate and that, provided the mass was less than 1.4 Suns (now known as the Chandrasekhar limit), such a body could cool without further gravitational collapse. He was managing editor of *The Astrophysical Journal* for almost 20 years and recipient of numerous honours, notably a share of the Nobel Prize for physics in 1983.

SIR FRED HOYLE (1915–2001)

English astrophysicist and cosmologist. In 1948, with Hermann Bondi and Thomas Gold, Hoyle proposed the Steady State theory of the Universe. Hoyle was also responsible, with Margaret (b. 1919) and Geoffrey Burbidge (b. 1925) and William Fowler (1911–95), for the theory of nucleosynthesis, the production of the heavy chemical elements in nuclear reactions in stars. The theory – known by the acronym B^2FH after the initials of the four – applied nuclear physics to astrophysics, and argued that the synthesis of the elements found on Earth was the result of a series of nuclear reactions which occur in supernova explosions and the final stages of evolution of red giant stars.

KEY PEOPLE: SCIENTISTS

SCIENTISTS

ALHAZEN (c. AD 965–1039)

Arabic scientist. Alhazen, building on Aristotle's *Meteorologia*, did fundamental work in optics. Dissecting the eyes of cattle to reveal their optical structures, he went on to project images with a camera obscura, make simple lenses and study the colours into which white light broke down when refracted. Despite this, it appears that Alhazen never questioned Aristotle's premise that light coming from astronomical bodies was pure and white, and decomposed into six colours only when contaminated by a passage through air, water or glass – an idea later overthrown by Isaac Newton (1642–1727). Alhazen calculated that atmospheric refraction causes sunlight to scatter in the air when the Sun is 19° below the horizon. This led him to question the density of the atmosphere and the altitudes of clouds that reflected sunlight while the Earth's surface was still dark.

ISAAC NEWTON (1642–1727)

English scientist. A great mathematical genius, Newton was also an experimental scientist, as his work on optics testifies. He was elected to the Royal Society in 1672, and after solving the gravity problem, in *Principia* (1687), he rapidly won international renown and became the archetype of a scientific genius. Taking the positions of Master of the Mint and President of the Royal Society, he was at the time the most powerful figure in British science, although he remained difficult to deal with throughout his life.

))))➡ *Newtonian Cosmology, p18*

JOHN HARRISON (1693–1776)

English chronometer maker and carpenter from Yorkshire, who developed a chronometer, or sea clock, capable of keeping time to the one minute of error per month allowed by the Longitude Act of 1714. This was a huge task, for Harrison had to design mechanisms to compensate for temperature changes and the rolling of the ship. By 1761, he had produced four chronometers, each one an improvement on the one before, and while the Admiralty backed the cheaper lunar

method, Harrison eventually earned himself a £20,000 prize in 1773 for inventing an accurate timekeeping mechanism for use at sea.

BELOW: Ernest Rutherford made numerous contributions to science, notably his demonstration of the existence of the atomic nucleus.

ERNEST RUTHERFORD (1871–1937)

New Zealand-born scientist. One of the most influential figures of twentieth-century science, by the time Rutherford had demonstrated the existence of the atomic nucleus, through his study of alpha-particle scattering, he had already won a Nobel Prize for chemistry (in 1908 for his work on radioactivity). Through his work on nuclear reactions he was the first scientist to realize the alchemists' dream of transmutation of the elements (changing one element into another).

Rutherford's greatest mistake was the belief that nuclear power could never be realized. He famously described as 'moonshine' the notion that humanity could ever harness the energy trapped within atomic nuclei. He died in 1937, two years before the discovery of nuclear fission.

HENRY NORRIS RUSSELL (1877–1957)

American scientist. Russell made major contributions to laboratory and theoretical spectroscopy and to stellar astrophysics, interests he combined to make the first reliable determination of the abundances of the elements in the Universe. His pioneering work on stellar evolution, starting with the Hertzsprung-Russell diagram, was followed by equally seminal work on eclipsing binary stars, which provides one of the most fertile testing grounds for theories of stellar structure.

)))**➤** *Hertzsprung-Russell Diagram, p115*

BERNHARD SCHMIDT (1879–1935)

Estonian telescope-maker. Schmidt designed a wide-field telescope that made possible large-scale photographic surveys of the night sky. One drawback of conventional reflecting telescopes is that they produce sharp images over only a narrow patch of sky. Schmidt's idea was to place a thin glass corrector plate in front of a deeply curved spherical mirror. The glass was subtly shaped to correct the distortions that the mirror would produce, resulting in a distortion-free image over several degrees of sky.

Schmidt produced his first telescope in 1930 and large Schmidt cameras with apertures of a metre or more have been in routine use in photographic sky surveys.

ARTHUR C. CLARKE (b. 1917)

British science writer and science-fiction author. The first person to propose geostationary communications satellites, Clarke's novels often reveal a realistic appreciation of orbital mechanics.

After working on wartime

RIGHT: Henry Norris Russell, along with Ejnar Hertzsprung, revolutionized the study of stellar evolution.

radar, Clarke published an article in *Wireless World* in 1945 in which he foresaw a global communications network, including worldwide television, linked by relay stations in geosynchronous orbits. This was 12 years before the first satellite was launched, 20 years before the first geosynchronous communications satellite (Early Bird), and more than 40 years before direct broadcast satellites fulfilled his prediction of beaming television straight into people's homes.

EUGENE SHOEMAKER (1928–97)

American lunar geologist. Although best known for co-discovering (with David Levy) the Shoemaker-Levy 9 comet that hit Jupiter in July 1994, Shoemaker's signature work was his research on the nature and origin of the Barringer Meteor Crater in Arizona, which helped provide a foundation for research into cratering on the Moon and planets.

Shoemaker was an inspiration behind America's lunar exploration. He took part in the Ranger lunar robotic missions, was principal investigator for the television experiment on the Surveyor lunar landers, and led the geology field investigations team for the first Apollo landings. He later conceived the 1994 Clemetine lunar mission. In 1997 he was killed in a car crash.

CARL SAGAN (1934–96)

American scientist. As a student in the 1950s Sagan's doctoral thesis included a section on 'The Radiation Balance on Venus'. Great minds such as Harold Urey (1893–1981), Fred Whipple (b. 1906), and Fred Hoyle (b. 1915), all disagreed about what existed under Venus's thick cloak of clouds. The confusion only increased when observations told of a blisteringly high temperature. Sagan's thesis, building on the work of those before him, pointed to a greenhouse effect as the logical culprit, a result that was confirmed when Mariner 2 flew past Venus in 1962. One of the outstanding popularizers of astronomy, Sagan wrote many best-selling books, and his television series, 'Cosmos', drew the largest audiences ever for a public television series.

ROCKET SCIENTISTS

KONSTANTIN TSIOLKOVSKY (1857–1935)

Russian rocket scientist. This visionary prophet of rocket travel published an explanation of how rockets could fly in the vacuum of space in 1883, and in *Dream of the Earth* (1895), he predicted artificial satellites positioned 300 km (190 miles) from Earth.

In 1903 selected chapters of his thesis, *The Exploration of Space Using Reaction Devices*, described multi-stage, liquid-fuelled rockets, stabilized by gyroscopes and steered by tilting rocket nozzles. The spacecraft was tear-shaped, with a passenger cabin in the rocket nose housing life-support systems and protected from extremes of temperature and the threat of meteoroids by a double skin. Tsiolkovsky even described details of the missions such as escape velocity, re-entry and the idea of a spacewalk, although he prophesied that these would not happen until the twenty-first century. Alexei Leonov's (b. 1934) first spacewalk was only 62 years later. In his 78 years, Tsiolkovsky never built a rocket, preferring to write and sketch his ideas.

ROBERT GODDARD (1882–1945)

American scientist who pioneered the use of liquid fuels to power rockets. In 1915 he was the first to show experimentally that the rocket could work in a vacuum, and hence outer space. His key publication, *A Method of Reaching Extreme Altitudes* (1919), pointed out the possibility of sending a small rocket to the Moon. In 1921, recognizing that liquid fuels carried more energy potential, Goddard began to tackle problems such as fuel injection, ignition and engine cooling. On 16 March 1926, Goddard launched the world's first liquid-propelled rocket. In a flight lasting 2.5 seconds, the liquid oxygen- and gasoline-powered machine flew just over 60 m (200 ft), reaching a height of 13 m (43 ft) at an average speed of 96 km/h (60 mi/h). Goddard continued to refine his liquid-fuelled rockets, reaching a maximum height of 2,530 m (8,300 ft) and a speed of 500 km/h (300 mi/h) with his A-series liquid rockets.

SERGEI PAVLOVICH KOROLEV (1906–66)

Russian rocket scientist. Korolev was an active member of the Moscow Group for the Study of Rocket Propulsion, with whom he helped to develop the first liquid-fuelled rockets launched in the USSR. Imprisoned by Stalin in 1938, he was forced to work in a scientific labour camp during the Second World War. Once freed, Korolev returned to rockets, working to improve the design of the captured V-2 missile and turning it into the first Soviet intercontinental ballistic missile, which carried the Sputnik satellites into orbit. Korolev later went on to mastermind the development of manned Vostok and Voskhod spacecraft and a series of robotic missions to the near planets.

WERNHER VON BRAUN (1912–77)

German rocket scientist who advocated the use of rockets for spaceflight. Despite the success of the liquid rockets he designed for the German army during the Second World War, von Braun saw them as space vehicles and designed larger rockets capable of orbiting a 30-tonne payload. After the War, he played a major role in developing Saturn 5, which took the Apollo astronauts to the Moon. Over 15 years of testing and flying for the US Army turned his V-2 into the rockets that would launch America's first artificial satellite and take the first American into space.

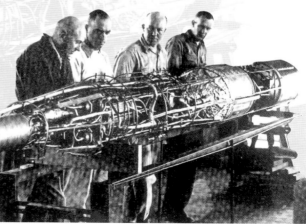

RIGHT: Robert Goddard (left), the pioneer of liquid-fuelled rocketry in the US, checking one of his most sophisticated models with his team in 1940.

MATHEMATICIANS

PIERRE SIMON, MARQUIS DE LAPLACE (1749–1827)

French mathematician and astronomer. In 1785 he was recognized as France's leading mathematician and between 1799 and 1825 he published the five huge volumes of his *Mécanique Céleste* ('*Celestial Mechanics*'), thought to be the finest work on the subject since Newton's *Principia Mathematica* (1687). Laplace believed that though God had created the Universe, it was now an ordered, self-running system, a belief which heralded the Age of Reason.

In an earlier book, *Système du Monde*, published in 1796, Laplace, noted that if a star was big enough or heavy enough, the force of gravity acting on light particles would prevent them from escaping. As Laplace put it, 'the attractive force of a heavenly body could be so large that light could not flow from it'. Although his reasoning was wrong – Laplace assumed light consisted of particles which would be influenced by gravity – and he did not use the term, Laplace had defined the black hole.

JOHN COUCH ADAMS (1819–92)

English mathematician. By 1845, from his analysis of the motion of Uranus, Adams had worked out an approximate position for an unseen more distant planet. French astronomer Urbain Le Verrier (1811–77) arrived at a similar conclusion in 1846. Using LeVerrier's calculations, German astronomers Johann Galle (1812–1910) and Heinrich D'Arrest (1822–75) found Neptune on 23 September 1846. Today both LeVerrier and Adams are credited with the discovery of Neptune.

ALEKSANDR FRIEDMANN (1888–1925)

Russian mathematician. Friedmann developed cosmological models based on general relativity.

He saw cosmology as an exercise in mathematics and did not attempt to relate it to the physical Universe. It is said that he only became interested in cosmological models after noticing an error in Einstein's 1916 work, which Einstein first disputed and then later accepted. Today, cosmological models with a cosmological constant of zero are termed Friedmann universes.

SIR JAMES JEANS (1877–1946)

British mathematician and physicist. In 1902 he showed that a cloud could only condense under gravity if it met certain initial conditions of density, size and temperature. His findings remain the basis for all modern work on the formation of stars and galaxies. Rejecting the idea that the Solar System had condensed in a similar fashion, he preferred the view that the material which formed the planets had been drawn out of the Sun by the gravitational pull of a passing star.

In 1928, 20 years before the development of the Steady State theory, he proposed that matter was continually created in the centres of spiral galaxies.

)))▶ *Steady State Theory, p48; Spiral Galaxies, p150*

MICHAEL MINOVITCH (b. 1936)

American mathematician. While working as a vacation student at the Jet Propulsion Laboratory (JPL) in 1962, Minovitch independently discovered a solution to the problem of how to accelerate and redirect a spacecraft by means of a close encounter with a planet. Although the English scientist Derek Lawden is credited with first solving the basic problem in 1954, it was Minovitch's work that brought this 'sling-shot' technique to prominence in the US.

)))▶ *Gary Flandro, p34*

ABOVE: A disk of gas and dust surrounding a black hole.

3,000 K, since any atom that did form would be broken up again by collisions. Before this time the Universe was a sea of atomic nuclei and electrons – a plasma rather than a gas. This plasma was opaque, radiation being continually emitted and re-absorbed so that matter and radiation were at the same temperature. When the Universe cooled below this critical temperature, stable atoms could form and so the link between matter and radiation was broken – they were 'decoupled' – and the two evolved independently of each other. Space became transparent for the first time and radiation could pass moreorless freely through the Universe. For this reason the period during which atoms formed is sometimes known as the decoupling era, although it is more often referred to as the recombination era (a slightly inappropriate name since nuclei and electrons were combining to form atoms for the first time rather than recombining).

Beyond Recombination

Since the recombination era, the cosmic background radiation has passed through the Universe without interacting with matter. Before recombination the

Universe was opaque, so the background radiation provides us with our earliest view of it. The temperature of the radiation is now 3 K, and since it was 3,000 K at recombination we know the Universe has expanded by a factor of 1,000 since then. With current cosmological models this places the start of recombination at a Universe age of about 300,000 years. The most distant galaxies that can be detected with present-day telescopes are seen at a time when the Universe was about 1,000 million years old. We have no observations of the Universe when its age lay between one million years and 1,000 million years, and the evolution of the Universe in this period, including the era of galaxy formation, remains very uncertain.

)))) ➤ *Hubble's Constant, p19; Planck Era, p46; Inflationary Era, p47; Hubble's Law, p47; Hubble Time, p47; Steady State Theory, p48*

ABOVE: German-born Physicist Albert Einstein (1879–1955) was responsible for the famous equation $E = mc^2$.

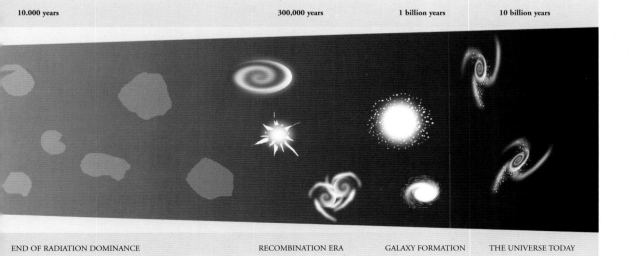

| 10.000 years | 300,000 years | 1 billion years | 10 billion years |

| END OF RADIATION DOMINANCE | RECOMBINATION ERA | GALAXY FORMATION | THE UNIVERSE TODAY |

PLANCK ERA

A very early phase of the Universe covering the time to when the Universe was 10^{-43} seconds old. Before this Planck time, general relativity and quantum mechanics come into conflict, and our present laws of physics break down. There is, as yet, no satisfactory quantum theory of gravity that would guide us in understanding the development of the Universe up to the Planck time.

BIG CRUNCH

Until recently, astronomers believed that the future of the Universe depended on its total density. If this is greater than a so-called 'critical density', then some time in the future it will stop expanding and start to contract, eventually collapsing into a hot, dense state called the Big Crunch. If the total density is less than the critical density, it will continue to expand forever. If it is the same, then the expansion rate of the Universe will slow down. In 1999, however, observations indicated that the expansion rate is accelerating. Astronomers do not yet fully understand the reason for this, but if they are correct, there will never be a Big Crunch.

ANTHROPIC PRINCIPLE

The anthropic principle attempts to explain the values of the fundamental constants by arguing that they are the only ones consistent with the evolution of intelligent life in the Universe. It also provides an explanation of several coincidences which are found in Nature such as the 'resonance' occurring in the nuclear reactions in which the element carbon is synthesized in stars. The anthropic principle also provides answers to fundamental questions which might normally be considered outside the realm of physics.

Not all physicists are happy with the use of the anthropic principle to answer such questions. Although the principle does not argue that the Universe was made for the benefit of mankind, it does give human beings, the observers, a special role in the Universe and, to some physicists, this is counter to the objectivity of physics.

CAUSALITY

Causality is the principle that an effect can follow a cause, but cannot precede it. As a scientific concept it has far-reaching consequences, not least because ideas which violate causality are always rejected by scientists. The theory of special relativity predicts that the time interval between two events will be different for two observers if they are in relative motion, the difference they perceive depending upon their relative speeds – the faster they are travelling relative to each other the greater the difference. But there will be no change in the order of the events. If there were such a change then causality would be violated for one of the observers.

)))》 *Special Theory of Relativity, p18*

HORIZON PROBLEM

When we measure the cosmic background radiation from any direction we are looking at gas which is so far away that the radiation from it has taken almost the age of the Universe to reach us. Looking in the opposite direction, we find that the same is true. Obviously no radiation can have passed between these two regions to equalize their temperatures and so physical contact between the two appears to violate the principle of causality. This is the 'horizon problem'.

)))》 *Inflationary Era, p47*

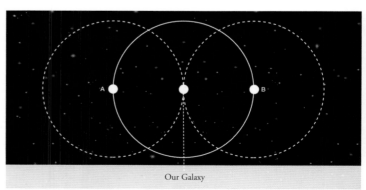

BELOW: 'Horizon Problem'. The solid circle drawn around our Galaxy represents our cosmic horizon. This is the limit to the distance we can see, set by how far light can have travelled during the lifetime of the Universe. No radiation can have passed between points A and B, on opposite sides of the horizon.

Our Galaxy

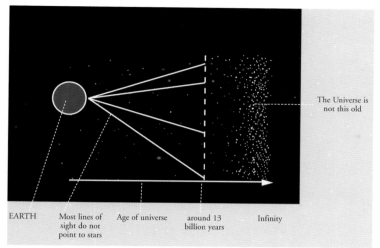

EARTH | Most lines of sight do not point to stars | Age of universe | around 13 billion years | Infinity

The Universe is not this old

$V_r = H_0 d$ – where the value H_0 is known as Hubble's constant.

HUBBLE TIME

From Hubble's law, if the expansion of the Universe has been steady, then the time which has elapsed since the Big Bang can, in principle, be calculated by using $1/H_0$ where H_0 is Hubble's constant. This is often referred to as the Hubble time. If we know Hubble's constant we also know the Hubble time. The current best estimate of the Hubble time is 13–14 billion years.

))))▶ *Hubble's Constant, p19*

INFLATIONARY ERA

A solution to the horizon problem was proposed by Alan Guth. He suggested that during its earliest stages – between roughly 10^{-34} s and 10^{-32} s – the Universe underwent a period of extremely rapid expansion, during which the expansion rate was actually accelerating. This period is known as the inflationary era (often shortened to 'inflation'), and during it the Universe's size increased dramatically. Although this solution may appear contrived, inflation simultaneously explains several features of our observed Universe.

))))▶ *Horizon Problem, p46*

HUBBLE'S LAW

Edwin Hubble (1889–1953) noticed that while a few bright galaxies had blueshifts, the fainter galaxies had only redshifts. He also noticed that galaxies with the largest redshifts tended to have the smallest images, and so were farther away. Using this limited data, he proposed that the galaxies were receding from us at a rate that was proportional to their distance. The implication of this was profound – at some finite time in the past the whole observable Universe must have been concentrated into a very small region, i.e. the Universe had a beginning.

Hubble's law states that the recessional velocity, V_r, of any distant galaxy is proportional to its distance, d, and is normally expressed as a simple equation:

OLBERS' PARADOX

Why is the sky dark at night? This problem is known as Olbers' paradox, after the German astronomer Heinrich Olbers (1758–1840). Although stars are far away, if we draw a line of sight outwards from Earth in an infinite Universe, every line would eventually reach a star. Since stars are similar to the Sun, the night sky would appear as bright as the surface of the Sun and given enough time the Universe would heat up to the surface temperature of the Sun. Olbers thought that the explanation for the dark night sky was that light from distant stars was absorbed by interstellar material such as dust clouds. But given enough time, the absorbing material would itself heat up and become luminous.

The night sky remains dark because the Universe has a finite lifetime. Because the speed of light is finite, we can see no farther than light can travel in the age of the Universe. Whether or not the Universe as a whole is infinite, the observable Universe is always finite and bounded. Stars beyond that distance cannot be seen because their light has not had time to reach us.

))))▶ *Hubble Time, above*

ABOVE: The resolution to Olber's Paradox lies in the fact that light from the most distant stars cannot have travelled for longer than the age of the Universe – between 13 and 15 billion years – and that light from an infinite number of stars could not have reached Earth.

SPACE-TIME

Einstein's special theory of relativity presented a new view of the nature of time and space. In Newtonian mechanics there are three dimensions of space and one of time. Any event can be located by specifying where it happened (three co-ordinates) and when (one co-ordinate). But while for Newton space and time exist independently of each other, for Einstein they do not. In Einstein's universe space and time are entwined in a four-dimensional entity called space-time. This means that changes in intervals of time for one observer are associated with changes in distance in space for another, but the separation between events in space-time is the same for both. It is in this sense that we live in a four-dimensional universe. It remains true, though, that space and time are different in kind. The equations of relativity treat space and time differently, but space and time are intimately interwoven.

))◗ *Special Theory of Relativity, p18*

WORMHOLE

A hypothetical structure that allows different regions of space-time to be connected by tunnel-like shortcuts. A concept originally implied by solutions to equations arising in Einstein's general theory of relativity, modern cosmology suggests that space-time has, on very small scales, a foam-like structure pervaded by wormholes.

COSMOLOGICAL CONSTANT

In order to overcome the problem of unstable static models of the Universe, Einstein introduced an extra term into his equations. He called it the cosmological constant. In essence the constant allowed for the existence of a very weak repulsive force, detectable only on the very large scales associated with cosmology, which would balance gravity and so allow a static Universe. Hubble's subsequent discovery of the recession of the galaxies removed the need for a static Universe and Einstein retracted his proposal for a cosmological constant. However some recent observations suggest the expansion of the Universe is accelerating and such an acceleration requires a repulsive force.

))◗ *BOOMERanG, p62*

COSMOLOGICAL PRINCIPLE

The view that there is nothing special about Earth, or the Sun, or even our own Galaxy has become enshrined in what is termed the 'cosmological principle'. This principle states that on the large scale the Universe is isotropic (it looks the same in all directions) and homogenous (it looks the same from all points). It follows directly from the principle of homogeneity that the Universe can have no edge. Despite the fact that the principle cannot, as yet, be justified by observation it has nevertheless played an important role in theoretical cosmology, as any cosmological model which does not conform to the principle is considered unsatisfactory.

))◗ *Steady State Theory, below*

STEADY STATE THEORY

In the 1950s it was argued that the cosmological principle ought to be extended so that the Universe should not only be homogeneous in space, but also that its large-scale properties should not change with time. To differentiate such an extension from the basic principle, the extension became known as the 'perfect cosmological principle', which led to the development of the 'Steady State' theory of the Universe, primarily by Austrian-born English physicist Hermann Bondi (b. 1919), the Austrian Thomas Gold (b. 1920) and English astronomer Fred Hoyle (1915–2001).

Since the expansion of space is moving galaxies apart and, therefore, reducing the overall density of the Universe, the perfect cosmological principle requires that new matter is continuously created in space, this matter eventually condensing to form galaxies and so maintaining a constant density of the Universe. The required rate of creation is surprisingly low, about one hydrogen atom per cubic metre of space every 10^{10} years. The proponents of the Steady State theory were also influenced by Mach's principle, which implies that the laws of physics would change in an evolving Universe. However, despite its attraction most cosmologists abandoned the Steady State theory after the discovery of the cosmic background radiation in 1965. The radiation was a natural consequence of the Big Bang, but could not be explained by the Steady State theory.

))◗ *Mach's Principle, p33; Hubbles's Law, p47*

SCHWARZSCHILD RADIUS

According to general relativity, if a body's gravity is sufficiently strong, space-time becomes so highly curved around the body that light is unable to escape from it. The German physicist Karl Schwarzschild (1873–1916) showed that this occurs when the radius of the body is less than a certain critical value, which increases with the body's mass. This critical radius is now known as the Schwarzschild radius. In Newtonian theory it equates to the distance at which the escape velocity from the body becomes equal to the speed of light, equivalent to the Laplacian view that light has been slowed and fallen back to the object.

DRAKE EQUATION

The Drake Equation, proposed by Frank Drake in 1961, identifies specific factors which might influence the development of technological civilizations and allows us to estimate how many might exist.

$$N = R^* \times f_p \times n_e \times f_i \times f_i \times f_c \times L$$

N = Number of communicative civilizations in the Milky Way whose radio emissions are detectable.

R^* = Rate of formation of stars.

f_p = Fraction of sun-like stars with planets; this is currently unknown, but seems to grow with discoveries made.

n_e = Number of planets in the stars habitability zone for each planetary system.

f_i = Fraction of planets in the habitability zone where life develops.

f_i = Fraction of life-bearing planets where intelligence develops. Life on Earth began over 3.5 billion years ago. Intelligence took a long time to develop. On other life-bearing planets it may happen faster, take longer, or not develop at all.

f_c = Fraction of planets where technology develops, releasing detectable signs of existence.

L = Length of time such civilizations release detectable signals into space. The number of detectable civilizations depends strongly on L. If L is large, so is the number of signals we might detect.

HUBBLE DEEP FIELD SURVEY

The deep field survey carried out by the Hubble Space Telescope in 1995 revealed galaxies much further away than any previously studied. Over 300 exposures were made in the direction of the north galactic pole, each with an exposure time of 15–40 minutes and in four colours to cover the electromagnetic spectrum. More than 1,500 separate galaxies were identified. A large number of faint blue galaxies were observed which were present in the early Universe but are not present now. Present elliptical galaxies were perhaps formed from mergers of these early, smaller galaxies. In 1998, a similar study was carried out in the southern celestial hemisphere – the Hubble Deep Field South. Its findings confirmed the theory that the Universe is uniform in all directions. In 2004, an even deeper observation was made, called the Hubble Ultra Deep Field.

ABOVE: The sky as seen in the Hubble Deep Field (HDF) observation, the deepest-ever view of the distant Universe, like a cosmological core sample. These views capture an assortment of galaxies at various stages of evolution, some dating back to within a billion years of the Big Bang.

MATTER AND PHENOMENA

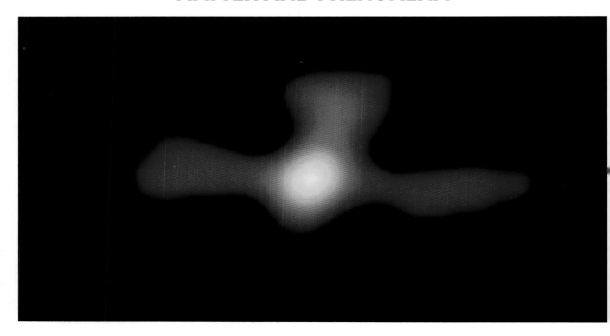

ANTIMATTER

Conservation laws dictate that antiparticles are produced every time a fundamental particle is created. Antiparticles have exactly the same characteristics as their particles except they have the opposite electric charge. Which we call 'matter' and which 'antimatter' is arbitrary. The creation of antiparticles would have created a Universe consisting of half matter and half antimatter. If antimatter does exist in the Universe then it must be very well separated from matter or the two would annihilate. However, current theories mostly assume that there was a slight excess of matter over antimatter in the very early Universe and that the mutual annihilation of particles and antiparticles resulted in the Universe being dominated by the surviving residue of matter.

RIGHT: The annihilation of a proton and antiproton produces an unstable pion which decays into gamma rays and neutrinos.
ABOVE: Small-scale annihilation of matter and antimatter occurs in our Galaxy between electrons and positrons, producing a faint glow of gamma rays mapped here by the Compton Gamma Ray Observatory.

ANNIHILATION

The annihilation process is the reverse of the creation process, which brought the matter of the Universe into existence and is subject to the same conservation laws. This means that a particle can annihilate only with its own antiparticle. When an electron and antielectron (otherwise known as a positron) annihilate, their masses are converted into gamma rays. When a proton annihilates with an antiproton their mass is converted into pions, which rapidly decay into gamma rays and neutrinos. Annihilation of matter and antimatter would therefore create large fluxes of gamma rays.
⫸ *Gamma Rays, p59*

HAWKING RADIATION

In 1974, theoretical physicist Stephen Hawking showed that a black hole, even without an accretion disk, will emit thermal radiation. This represents a loss of energy, and therefore mass, so the black hole will slowly evaporate, eventually disappearing in a burst of radiation. According to quantum mechanics, space is not an empty vacuum in the conventional sense but is a 'soup' of so-called virtual particles which are continually popping in and out of existence. The virtual particles are created as pairs – one a particle and the other an antiparticle – and if such a pair is created just outside the event horizon of a black hole, one may be captured by the black hole while the other escapes. The separation process extracts energy from the gravitational field and this may be sufficient to convert the virtual particles into real particles. These pairs are continuously annihilating and producing Hawking radiation.

))))➤ *Antimatter, p50; Accretion Disk, p44*

ATOMIC NUCLEUS

The central part of an atom consisting of positively charged protons and uncharged neutrons, thus having an overall positive charge. The nucleus contains most of the mass of an atom.

PROTON

A fundamental particle that is a constituent part of every atom. Protons are part of the nucleus with a positive charge equal to that of an electron (1.602 x 10^{-19} coulomb) and a mass of 1.6726 x 10^{-27} kg.

NEUTRON

A fundamental particle that is a constituent part of all atoms except those of common hydrogen. Neutrons are part of the nucleus with no charge and a mass of 1.6749 x 10^{-27} kg.

ELECTRON

A fundamental particle that is a constituent part of every atom. An electron orbiting an atomic nucleus may occupy any one of a series of distinct energy levels dictated by quantum mechanics. Each electron has a negative charge of 1.602 x 10^{-19} coulomb and a mass of 9.109 x 10^{-31} kg.

BARYON

A class of elementary particles that have a mass greater than or equal to that of a proton (1.6726 x 10^{-27} kg), participate in strong interactions and have a spin of one half.

PLASMA

At very high temperatures gases become plasmas in which the atoms are broken down into negatively-

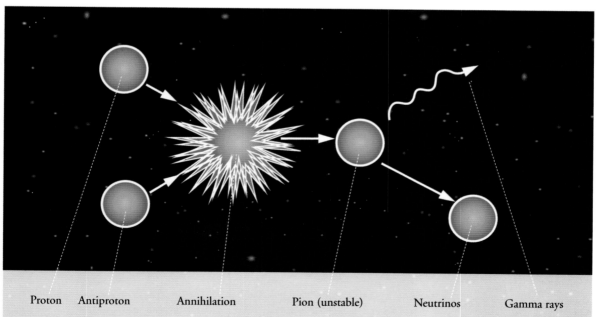

Proton   Antiproton   Annihilation   Pion (unstable)   Neutrinos   Gamma rays

charged electrons and positively-charged nuclei. The attraction between these charged particles gives a plasma properties which are different from those of normal gases. For example, various forms of wave motion can arise in a plasma and these can be excited by the absorption of radiation. Consequently plasmas are opaque over a wide range of frequencies, unlike normal gases which absorb at specific frequencies, the so-called spectral lines.

NEUTRINOS

Neutrinos are tiny, fundamental particles that are emitted in certain kinds of nuclear reaction. They possess little or no mass, interact extremely weakly with other matter and always move at, or very close to, the speed of light. Their interest to astronomers is that they are generated in the nuclear reactions that occur in the cores of stars like the Sun. They stream unhindered out of the Sun and pass straight through the Earth. Several large detectors have been constructed to search for solar neutrinos. In one type, chlorine atoms are turned into argon on the very rare occasions on which they collide with a neutrino. In another, gallium atoms are turned into germanium. In a third type, neutrinos occasionally collide with water molecules to produce flashe sof light. Other possible souces of neutrinos are supernova explosions (neutrinos from Supernova 1987A were detected by chance) and ative galaxies. Neutrinos created in the Big Bang may form part of the dark matter that cosmologists believe accounts for a large part of the Universe.

)))➤ **Big Bang, p44; Solar Neutrinos, p70; Supernovae, p126; Active Galaxies, p150**

COSMIC RAYS

Cosmic rays are highly energetic charged particles that constantly bombard the Earth from space. Their energies range from 10^8 to 10^{19} eV. The majority are protons (hydrogen nuclei) and most of the remainder are alpha particles (helium nuclei), with a few per cent being electrons. These primary cosmic rays collide with molecules in the atmosphere to produce large numbers of other particles (secondary cosmic rays) which in turn decay to produce an extensive air shower covering several square kilometres of the Earth. A single energetic primary particle can produce up to half a million secondary particles.

BELOW: The Egg Nebula, a cloud of dust and gas ejected from a former red giant, the core of which is hidden by a cocoon of dust (the dark band at the centre). The arcs are denser shells which arise from variations in the rate of mass ejection from the central star on time scales of 100 to 500 years.

EXOPLANETS

Extrasolar planets or exoplanets are planets around other stars. Most of the first exoplanets to be discovered were the size of Jupiter or larger and orbited closer to their parent star than Mercury does to the Sun because they are the planets that are easiest to spot with the radial velocity technique used. Such planets, known as 'hot Jupiters', may have formed farther out and then moved inwards as they were slowed by drag from the gas in the surrounding disk. Many of the exoplanets that lie farther away from their parent stars have orbits that are highly elliptical, unlike the near-circular orbits of the planets in our Solar System. Not until many more exoplanets are known can we tell whether our own Solar System is unusual. However, we can already draw important conclusions, notably that planets are common in the Galaxy and that perhaps the majority of stars have planets orbiting them. Several cases of multiple planets orbiting a single star have been found, confirming that entire planetary systems can arise. Planets tend to be found only around stars that, like our own Sun, are particularly rich in heavy elements. Most of the planets known have masses no more than a few times that of Jupiter, and as searches continue the number of lower-mass planets is increasing.

NEBULAE

Astronomers have traditionally used the term nebula (Latin for 'mist') to refer to any diffuse patch of light in the night sky. As early as 1781 Charles Messier (1730–1817) had drawn up a catalogue of about 100 nebulous-looking objects. With improved telescopes many of the nebulae proved to be clusters of stars, but others remained enigmatic. For many years a Great Debate raged over the nature of these misty objects. Some astronomers argued that they were clouds of gas between the stars of the Milky Way, while others maintained that they were huge star systems far outside our Galaxy. Interestingly, both views are now known to be correct. Some objects, such as the Orion Nebula, are indeed gas clouds within the Milky Way (and in modern usage a 'nebula' is such a cloud), but others are distant galaxies, far removed from our own.

)))➤ *Messier Catalogue, p56; Orion, p137; Milky Way, p154*

EMISSION NEBULA

A cloud of interstellar dust and gas that shines by its own light. The nebula is composed primarily of hydrogen that has been ionized. There are three main types of emission nebula: HII regions, planetary nebulae and supernova remnants.

)))➤ *Planetary Nebula, p122*

SPICULES

Short-lived (around 5 to 10 minutes) narrow jets of solar material on the chromosphere, seen at the limb (the edge of the visible disc of the sun). They have temperatures in the region of 10,000 to 20,000 K and velocities of 20 to 30 km/s. They are located at the edges of supergranulation cells.

)))➤ *Chromosphere, p71*

PRIMORDIAL GAS

Stars condense out of the interstellar gas in a galaxy and the primordial gas is that from which the first stars condensed. The present interstellar gas has been enriched with new elements synthesized in earlier generations of stars, so to learn about the make-up of the primordial gas we must look for members of the first generation of stars to be formed. Such first-generation stars are found in globular clusters. These stars are composed of 75 per cent hydrogen and 25 per cent helium, with small traces of lithium. If these elements existed before the first stars were formed they must have been created in the high temperatures of the early Universe.

)))➤ *Interstellar Medium, p149*

COSMIC BACKGROUND RADIATION

The discovery in 1965 of the cosmic background radiation was an accident. Arno Penzias and Robert Wilson were working at the Bell Telephone Laboratories in New Jersey on the development of receivers for communication satellites. Investigating the problem of radio noise, they had found that the sky was unexpectedly bright at a wavelength of 7.3 cm (2.9 in), in the microwave region of the radio spectrum. They concluded that the noise from all over the sky corresponded to the glow of a black body at a temperature of about 3 K.

A few miles away at Princeton University, Robert Dicke and Jim Peebles had realized that radiation from the hot gas that filled the Universe in the first moments after the Big Bang should still be visible, but redshifted by the Hubble expansion from visible light to very weak radio waves. It would have a black-body spectrum at about 10 K. They had begun an independent search for this 'remnant' radiation in the microwave part of the radio spectrum when they heard of the discovery of Penzias and Wilson. Neither group knew that the radiation they had found had been predicted 20 years earlier by George Gamow.

The initial measurements of Penzias and Wilson indicated that the cosmic background radiation was isotropic – uniform across the whole sky. This implied that the radiation originated before galaxies had had time to form, in a Universe consisting only of a primordial gas of uniform density. However, theories of the origin of the Universe required that the density of the primordial gas could not be completely uniform. In order to form clusters of galaxies, small variations in density must have existed, and these would show up as very small variations in the brightness of the radiation. COBE confirmed the existence of changes in intensity across the sky. Although tiny, corresponding to differences of just a few parts in 100,000, the changes are consistent with the expected density differences within the primordial gas. The cosmic background radiation is regarded as confirmation that the Universe did begin in a hot Big Bang.

⫸ *Big Bang, p44; Hubble's Law, p47*

ELECTROMAGNETIC WAVES

Electromagnetic waves were predicted by Maxwell, who showed that varying electric and magnetic fields could sustain each other and travel through space at the speed of light. The spread of wavelengths is known as the electromagnetic spectrum. Electromagnetic waves are emitted by vibrating electrons and absorbed when they cause other electrons to vibrate in sympathy with them, or they may be emitted and absorbed in changes of the electronic state of atoms and molecules.

⫸ *Photon, Electromagnetic Spectrum, below;*
James Clerk Maxwell, p33

PHOTON

A discrete 'packet' of electromagnetic radiation travelling at the speed of light. Electromagnetic radiation travels as waves but it interacts with matter as if it were composed of particles (photons). The energy (E) of the photon is directly related to the frequency (ν) of the electromagnetic radiation: $E = h\nu$, where h is the Planck constant.

ELECTROMAGNETIC SPECTRUM

The electromagnetic spectrum is the range of possible frequencies and wavelengths of electromagnetic waves. While there is a continuous gradation in these properties through the spectrum, they are conventionally classified into a small number of ranges. Gamma rays, at the extreme short-wavelength end of the spectrum, have wavelengths shorter than 0.01 nm. X-rays have wavelengths between 0.01–10 nm. Ultraviolet radiation lies beyond the short-wavelength end of the visible band, between 10 and 400 nm. Visible light lies between 400 nm (violet) and 700 nm (red). Infrared

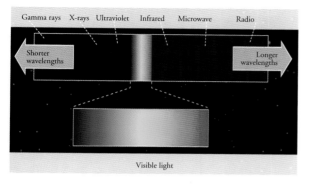

Gamma rays X-rays Ultraviolet Infrared Microwave Radio

Shorter wavelengths

Longer wavelengths

Visible light

ABOVE: Visible light occupies only a fraction of the electromagnetic spectrum. The bands of shorter wavelength radiation (ultraviolet, X-ray and gamma ray) and longer wavelength radiation (infrared, microwave and radio wave) are all studied in modern astronomy.

radiation lies beyond the long wavelength end of the visible spectrum and stretches between 700 nm and 1 mm. Radio waves extend beyond wavelengths of about 1 mm.
))))➤ *Gamma Rays, X-Rays, Ultraviolet Rays, Infrared Rays, Radio Waves, pp59–60*

GAMMA RAYS

Gamma rays were identified in 1900 by a French chemist, Paul Villard (1860–1934), but it was not until 1912 that they were recognized as the most energetic form of electromagnetic radiation, with extremely short wavelengths of less than 0.01 nm. In many ways they behave like high-energy X-rays, and there is no consensus on the wavelength at which X-rays end and gamma rays begin. A useful distinction is their source: gamma rays proper are emitted from the nuclei of atoms either during nuclear reactions or as a result of spontaneous radioactive decay. Gamma rays are detected by counting them one by one. In that sense, they behave more like particles than like waves.

X-RAYS

X-rays were discovered by the German physicist Wilhelm Röntgen (1845–1923) in 1895. Coming soon after Hertz's discovery of radio waves it was expected that they too would prove to be electromagnetic waves, but they seemed at first to show none of the characteristics of light, such as reflection and refraction, that also

characterized radio waves. It took until 1912 for physicists to discover that this was because their wavelength was extremely short, in the range 0.01–10 nm, many times shorter than light waves and comparable to the dimensions of atoms. X-rays are generated when high-speed electrons collide with a heavy metal target and give up their kinetic energy as electromagnetic radiation. Astrophysical sources of X-rays include plasmas with temperatures in the range of 10^6-10^8 K, and charged particles travelling at large fractions of the speed of light in magnetic fields.

ULTRAVIOLET RAYS

A German pharmacist, Johann Ritter (1776–1810), investigated the region beyond the blue end of the visible spectrum. In 1801 he discovered an invisible type of radiation that affected photographic plates and became known as ultraviolet radiation (UV). The UV has wavelengths in the range 10–400 nm. UV waves carry more energy than visible light does, and the higher-frequency radiation is able to knock electrons out of atoms (ionization), break chemical bonds and damage molecules. Ultraviolet radiation from the Sun is responsible for sunburn and certain types of skin cancer

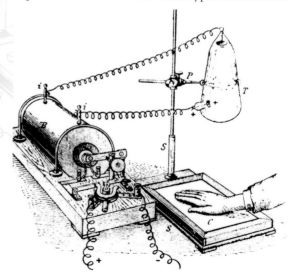

ABOVE: The first X-ray machine, pictured here, was invented by Wilhelm Röntgen in 1895.

RIGHT: From orbit the European Infrared Space Observatory (ISO) observed infrared radiation that does not penetrate Earth's atmosphere coming from objects such as regions of star formation, dying stars and galactic nuclei.

and is most dangerous in the 290–320 nm region known as UVB. Most of the UVB from the Sun is absorbed by atmospheric ozone. When UV rays strike certain materials the energy is absorbed and re-emitted in the form of visible light, a phenomenon known as fluorescence if the re-emission is immediate, or as phosphorescence if it persists after the stimulating UV ceases.

INFRARED RAYS

The infrared region of the electromagnetic spectrum lies to the long-wavelength side of visible light and was discovered in 1800 by the German-British astronomer William Herschel, who placed a thermometer beyond the red end of a spectrum of sunlight formed by a prism. The thermometer rose, indicating that it was being heated by radiation invisible to the eye. Infrared radiation, which extends from 700 nm to 1 mm, is sometimes known as 'radiant heat' because it is emitted by objects at normal temperatures and can be felt as warmth on the skin. This term can be misleading, since all forms of electromagnetic radiation, including visible light, can warm matter if they are absorbed. Radiation at infrared wavelengths is readily emitted and absorbed by vibrating or rotating molecules. Absorption by carbon dioxide and water molecules make the atmosphere opaque in most of the infrared band. These molecules are also largely responsible for the greenhouse effect. The infrared region of the spectrum is roughly divided into the near, mid and far infrared.

RADIO WAVES

The longest electromagnetic waves, from about 1 mm upwards, are known as radio waves. They were discovered by the German physicist Heinrich Hertz in 1888 while testing the predictions of Maxwell that electromagnetic waves should exist at longer wavelengths than those of light. Radio waves occupy the widest stretch of the electromagnetic spectrum and the region is often subdivided into standard bands for convenience. The shortest waves, less than about 10–30 cm, are called

microwaves. The relatively long wavelengths and low frequencies allow radio waves to be generated under precisely controlled conditions by causing an electric current to oscillate in a suitable circuit.

))⟩ *Heinrich Hertz, p33; James Clerk Maxwell, p33*

REDSHIFT

The increase in wavelength of electromagnetic radiation emitted from a source, due either to the source's own movement away from the observer (Doppler effect) or to the expansion of the Universe (cosmological redshift). The opposite effect is blueshift.

))⟩ *Doppler Effect, below*

DOPPLER EFFECT

The Doppler effect applies to any kind of wave motion and occurs whenever the observer is moving with respect to the source of the waves. There is a change in frequency of electromagnetic radiation that arises as a result of the relative motion between the source of radiation and the observer. When the source moves away from the observer the frequency is decreased (and the wavelength correspondingly increased) and any spectral lines produced by that source will be redshifted relative to the corresponding lines from a stationary source. Conversely, the observed frequency is increased (and the wavelength decreased, or 'blue-shifted') when the source is approaching the observer. For speeds that are small compared with the speed of light, the amount of change (the Doppler shift) of an electromagnetic wave is given by: $\lambda_1 = \lambda_0 (1 + v/c)$ – where λ_1 is the emitted wavelength, λ_0 is the observed wavelength, v is the relative speed of the source away from the observer, and c is the speed of light.

GLITCH

 A temporary speeding up in the rotation of a pulsar. Pulsars gradually slow down over time and these glitches are believed to be due to adjustments occurring in the core or crust of the neutron star.

⟫▶ *Pulsars, p128*

GRAVITATIONAL WAVES

Gravitational waves are ripples in space-time predicted by Einstein's general theory of relativity.

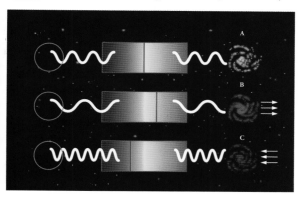

ABOVE: The Doppler effect changes the wavelengths of light from stars and galaxies in motion. Galaxy A is at a constant distance from Earth so the Doppler effect does not occur. Galaxy B is receding from Earth and the wavelengths of light from it are 'stretched', shifting them towards the red end of the spectrum ('redshifted'). Galaxy C is approaching Earth and the light is squeezed to shorter wavelengths ('blueshifted').

In the 1960s, experiments in the US appeared to show that gravitational waves were coming from the centre of the Galaxy. Other experiments failed to confirm this and it is now thought that the claims were mistaken. Several research groups are nonetheless constructing highly sensitive instruments which they hope will be the first to catch gravitational waves. The most ambitious project is the Laser Interferometer Gravitational Wave Observatory (LIGO) which is being built on two sites in Louisiana and Washington in the US. Each installation is an L-shaped structure with 4-km (2.5-mile) arms. Masses suspended at the corner of the 'L' and at the ends of the arms will move in response to a passing gravitational wave. Laser beams directed along the arms will sense these tiny motions of no more than one thousandth the diameter of a proton. Similar, though smaller-scale, observatories include VIRGO and GEO in Europe. An international proposal called LISA (Laser Interferometer Space Antenna) envisages a space-borne interferometer consisting of three spacecraft forming an equilateral triangle with sides measuring 5 million km (3 million miles) long. Lasers shining between the spacecraft will monitor their separation and so detect passing gravitational waves. Possible sources of gravitational waves include very close binaries, colliding neutron stars, and stars collapsing into black holes.

NUCLEOSYNTHESIS

An understanding of the synthesis of the lightest elements in the first few minutes after the Big Bang is considered to be one of the cornerstones of modern cosmology. The abundance and distribution of the nuclei of hydrogen, helium and lithium provides a detailed record of conditions in the early Universe. As the Universe cooled down, these light elements began to coalesce under the action of gravity to form stars. The synthesis of nuclei continues within stars as the lighter nuclei fuse to form heavier elements such as carbon and oxygen. Additional nucleosynthesis occurs in supernovae. The products ejected in supernovae spread out into the interstellar medium, where they can form the raw material for a subsequent generation of stars and planets. Indeed, all the heavier elements in our bodies were 'cooked' inside an earlier generation of stars.

MEASURING THE UNIVERSE

ABOVE RIGHT: The BOOMERanG detector which obtained measurements of the scale size of the variations in the cosmic background radiation in 1998.

BOOMERanG

BOOMERanG is a balloon-borne detector, flown from Antarctica, which measured the scale size of the variations in the cosmic background radiation. These measurements showed that the geometry of space is flat, which is what we would expect if the very early Universe had an inflationary phase. Similar results were obtained by another balloon-borne experiment called MAXIMA, which flew high above Texas.

Flat space implies the density of the Universe is equal to the critical density, but measurements show that the density falls far short of this, even when dark matter is included. This discrepancy would be removed if a cosmological

RIGHT: NASA'S Cosmic Background Explorer (COBE) satellite produced a microwavable map of the whole sky, eventually collecting four years of data. Cosmology theory indicates that the density fluctuations observed could ultimately be responsible for the formation of galaxies.

constant or some other form of 'dark energy' exists.

》》》 *Inflationary Era, p47; Cosmological Constant, p48; Dark Matter, p53*

COBE

The Cosmic Background Explorer (COBE) satellite was launched in 1989 to search for small variations in the cosmic background radiation. The results from the satellite confirmed that the spectrum of radiation had the shape of a black-body spectrum at a temperature of 2.725 K with a deviation of just one part in 100,000, but also that there were changes in intensity across the sky.

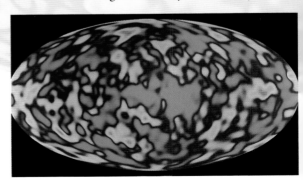

COMPTON GAMMA RAY OBSERVATORY

The Compton Gamma Ray Observatory (GRO) was launched in 1991 and re-entered Earth's atmosphere in 2000. GRO had four independent telescopes. The Burst and Transient Source Experiment (BATSE) monitored the whole sky for short bursts of gamma rays with energies between 50 keV and 600 keV. The Oriented Scintillation Spectrometer Experiment (OSSE) looked for the gamma-ray spectral lines expected from radioactive decay and electron-positron annihilation. The Imaging Compton Telescope (COMPTEL) surveyed

ABOVE: The Compton Gamma Ray Observatory, launched from the Space Shuttle Atlantis in 1991, increased the known number of gamma-ray sources tenfold before burning up in the Earth's atmosphere nine years later.

the sky at gamma-ray energies from 1 MeV to 30 MeV, while the Energetic Gamma Ray Experiment Telescope (EGRET) was a spark-chamber detector sensitive to gamma rays from 20 MeV to 30 GeV.

SPECTROSCOPY

Spectroscopy is the branch of physics concerned with the formation and interpretation of spectra. Instruments used for observing spectra are variously known as spectroscopes, spectrometers, or spectrographs. Spectroscopic instruments use either a prism or a diffraction grating to form a spectrum. A diffraction grating is more common, as there is a simple relationship between the wavelength of the light and the angle through which it is diffracted. It is also effective over a wider range of wavelengths than a prism. In astronomy,

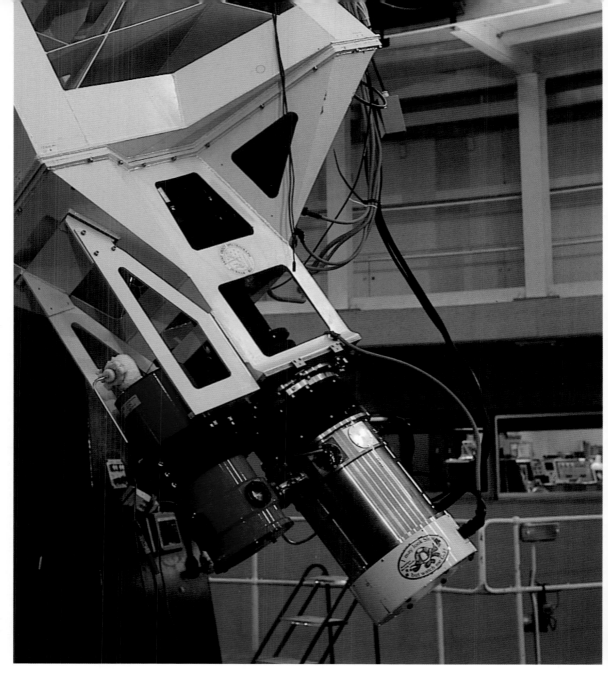

where almost all objects have to be observed remotely, spectroscopy is a powerful tool for studying the physical conditions in planets, stars, galaxies and interstellar space. Among its many applications, astronomers use spectroscopy to measure the chemical composition of stars and planetary atmospheres, the temperatures of gaseous nebulae, the strengths of magnetic fields in space, the rotation speeds of galaxies, the expansion rate of the Universe, and the masses of black holes.

ABOVE: A faint object Spectograph on the William Herschel Telescope. Such devices are used to create a photographic or electric image of a spectrum.

generation heating the gas, so creating pressure to expand, and the forces of gravity trying to crush the outer layers in towards the centre. At the surface, the temperature is only 5,500°C (9,900°F) and the density less than 0.001 kg/m³ (0.00006 lb/ft³). The mechanisms of energy transport change throughout the Sun: in the volume within 70 per cent of the radius, energy is moved outwards by radiation in the form of photons, repeatedly absorbed and re-emitted in a zigzagging 'drunkard's walk' – the radiation takes tens or even hundreds of thousands of years to reach the surface from the centre. For the final 30 per cent of the way, energy is transported by convection, with currents of hot gas rising to the surface to be replaced by descending cooler gas.

The Active Sun

The visible surface of the Sun, or photosphere, appears as a blindingly bright white disk. The telescope reveals that relatively dark areas, or sunspots, are usually present at the lower latitudes. The numbers of these grow and shrink in an 11-year cycle. At their maximum sunspots can occasionally be seen with the naked eye.

During a total solar eclipse, when the light of the photosphere is briefly cut off, the glowing red chromosphere, or lower atmosphere of the Sun, appears around the edge of the dark Moon. Extending beyond the Chromosphere is the pearly-white corona, or outer atmosphere, extending far into space. Astronomical instruments now make it possible to study these features at any time.

The solar cycle controls a host of other activity on and above the Sun's surface. Bright regions called *faculae* frequently appear before sunspots emerge; these are hotter and denser than surrounding material. Filaments of gas called prominences rise from the photosphere. Among the more violent events occurring on the Sun are flares, in which a huge amount of energy – the equivalent of as much as 10 billion one-megaton bombs – is released in a short time (typically minutes to hours). Very occasionally these can be seen in white light, but they also give rise to X-rays and ultraviolet emission, together with the ejection of energetic particles. Passing through the corona, these create shock waves and produce strong bursts of radio emission.

⫸ *Prominences, p70; Sunspots, p70; Main Sequence Stars, p131*

SOLAR WIND

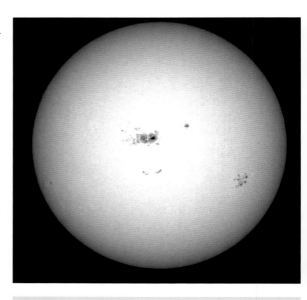

The continuous, though highly variable, outflow of charged particles (mainly electrons and protons) from the Sun.

SOLAR NEUTRINOS

Neutrinos produced by nuclear reactions in the heart of the Sun. Neutrinos are electrically neutral particles of extremely small – possibly zero – mass. They are elusive, being able to pass through Earth with only a minute chance of interacting with any atom on the way. Fewer neutrinos than originally predicted arrive at the Earth from the Sun. The mystery of the missing neutrinos was solved in 2001. They weren't missing at all, but on their way from the Sun to the earth they changed to different types of neutrinos that are harder to detect. When the total number of all the different types of neutrinos were counted, they agreed with predictions.

))))➤ *Neutrinos, p52*

ABOVE : *Dark sunspots on the Sun's surface.*

PHOTOSPHERE

The visible surface of the Sun. The photosphere has a temperature of about 5,500°C (9,900°F). Here the most common constituents, hydrogen and helium, are generally not ionized, although most of the heavier chemical species are at least partially so. This mixture of atoms and ions, plus some resilient, simple molecules, absorbs light emerging from the interior to produce a rich absorption-line spectrum in which the signatures of at least 65 chemical elements have been found. High-resolution photographs of the surface reveal a mottled appearance known as granulation, which changes on a timescale of 10 minutes or so. It is caused by rising and falling bubbles of gas. On a larger scale, supergranulation can be observed, in which vast convective cells, perhaps 30,000 km (20,000 miles) across, can be traced. The photosphere is marked by sunspots, regions that are relatively cool and dark compared with their surroundings.

))))➤ *Chromosphere, p71; Corona, p72; Granulation, Sunspots, below*

PROMINENCES

Filaments of gas extending above the photosphere, the visible disk of the Sun. They can often be seen during a total solar eclipse, fringing the dark disk of the Moon. Observations at hydrogen and calcium wavelengths show them some 40,000 km (25,000 miles) above the photosphere and extending typically for 100,000 km (60,000 miles). They show a range of structures, especially loops and arches. Some quiescent prominences may persist for months, while active ones last only a short time and can sometimes be seen rising from the Sun's surface, occasionally to heights of several hundred thousand kilometres.

GRANULATION

The small granular markings observable on the Sun's photosphere, or surface. They are the areas where hot gas is being brought to the surface by convection. Cells are around 1,000 km (600 miles) across and are part of a larger 'super-granulation' pattern.

SUNSPOTS

Relatively dark areas on the photosphere, or bright white surface of the Sun. They have a lower temperature than the rest of the photosphere, down to 3,700°C (6,700°F). Small spots, just a few hundred kilometres across, may last only a few hours, but some

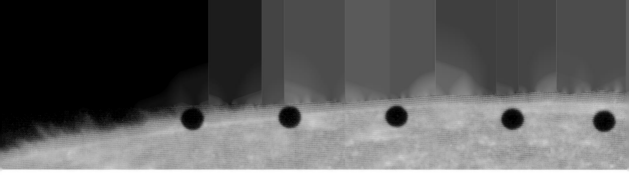

surprisingly, Sun-scorched Mercury may have polar caps composed of ice accumulated from cometary impacts.

Mercury's Past

Clues to the nature of Mercury's interior, and also its evolution, come from its unusually high density, second only to that of Earth. This is due to a large iron core that accounts for two-thirds of the planet's mass. Motions of liquid iron in this core give rise to a weak magnetic field, about 1 per cent the strength of Earth's.

Why should Mercury, a dwarf among planets, possess such a disproportionately large iron core? At birth, Mercury may have been twice its present size, but suffered a hit-and-run accident with a stray body of similar size to our Moon, which blasted off most of its less dense, outer rocky layers. This collision could also have been responsible for knocking Mercury's orbit into its current elliptical shape.

))))➤ *Caloris Basin, below*

CALORIS BASIN

Mercury's largest geological formation is a circular lowland 1,300 km (800 miles) wide, fully one-quarter of the planet's diameter. It is named the Caloris Basin (from the Latin *calor*, meaning 'heat') because it faces the Sun when Mercury is closest to the Sun. The basin was excavated by an immense impact.

Furnace and Freezer

Mercury has a highly elliptical orbit that takes it between 46 and 70 million km (29 and 43 million miles) from the Sun. Seen from Mercury's surface, the Sun appears from two to three times larger than it does from Earth, depending on whether Mercury is at its farthest (aphelion) or closest (perihelion). At perihelion, Mercury's daytime surface temperature can exceed 400°C (750°F) on the equator – hot enough to melt tin and lead.

At successive perihelia, first one side of the planet and then the other is presented to the Sun. The two points on the equator that face the Sun at perihelion, receiving the most intense solar heating, are sometimes termed the 'hot poles'. Yet, without an atmosphere to distribute heat around the planet, temperatures on Mercury's night side drop below -180°C (-292°F). Near the north and south geographical poles there may be permanently shaded regions where temperatures would remain sub-zero. So,

ABOVE: Occasionally, Mercury crosses in front of the Sun, where it can be seen in silhouette as a black dot, an event termed a transit. This series of images shows Mercury near the edge of the Sun during the transit of 15 November 1999, as seen by a Sun-watching spacecraft called TRACE. ABOVE LEFT: The Caloris basin is the largest formation on the surface of Mercury. This massive lowland plain, partly visible at lower left of the picture, was created by an asteroid collision, and measures 1,300 km (800 miles) across.

VENUS

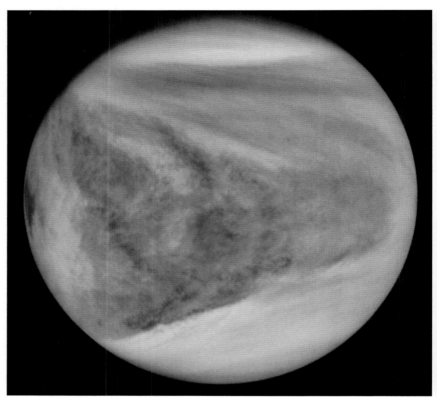

The second planet from the Sun and Earth's nearest neighbour. Venus has a mass and diameter similar to those of our own planet. Thick clouds constantly hide its surface. Once thought likely to be the most Earthlike of the planets, its true nature has been shown by visiting spacecraft: its dense, choking atmosphere makes the surface a hell of scorching heat and crushing pressure.

Seeing Venus

Venus moves within the Earth's orbit, and so never strays far from the Sun in our sky. Best seen shortly before sunrise or after sunset, it is commonly called the Morning Star or Evening Star. Its brilliance is a result of highly reflective clouds and its close proximity to Earth.

Because it moves inside Earth's orbit, Venus shows phases, like the Moon. These phases are apparent through binoculars or a small telescope.

The Surface of Venus

Underneath the clouds lies a tortured landscape, preserving the scars of geological processes that have been at work for hundreds of millions of years. Venus is peppered with volcanoes of all sizes, cut by a maze of faults and ridges, and plastered with sheet-like lava flows hundreds of kilometres wide.

Eighty per cent of the Venusian surface has a range in altitude of no more than 1,000 m (330 ft), making it much smoother than Earth. Venus also lacks plate tectonics – the constant movement and recycling of large crustal regions that has shaped the Earth's surface.

Although heights are less extreme, mountains and valleys do exist on Venus, testament to huge forces resulting from the shifting of molten material beneath the crust. The most complex features of the deformed terrain on Venus are *tesserae* (Greek: 'tiles'), high-standing plateaux characterized by a lattice of faults and fractures. *Tesserae* may be formed where Venusian crust is dragged and rucked up very much like a carpet, by the sinking of molten rock below.

Over 900 impact craters exist on Venus, compared with fewer than 200 on Earth. Almost all the craters appear fresh; very few are cut by faults or flooded by lava. The number and degree of preservation of Venusian craters give an indication of the average age of the

ABOVE: This view of Venus, taken from orbit by the Pioneer Venus orbiter, shows the swirling mass of thick cloud that obscures its surface.

RIGHT: Venus, also known as the Evening or Morning Star, is one of the sky's most easily visible objects to the naked eye.
BELOW: Volcanoes such as Maat Mons are ubiquitous on the surface of Venus and are an important way in which the planet loses internal heat.

surface, which is estimated to fall between 200 and 600 million years – much younger than the Moon's, but slightly older than Earth's.

Vulcanism

The surface of Venus is dominated by volcanic landforms. More than 80 per cent of the planet is covered with undulating lava plains. Tens of thousands of small volcanoes a few kilometres across litter the surface. Larger shield-style volcanoes, similar to the Hawaiian Island volcanoes on Earth, have produced aprons of complex lava flows hundreds of kilometres wide. In general, Venusian volcanoes are lower and much wider than those on Earth.

Coronae (Latin: 'crowns') are large, circular volcanic structures, typically hundreds of kilometres across, with collapsed centres. They appear unique to Venus. Each corona marks a 'blister' on the crust where heat and molten rock have welled up from below. Coronae, along with the hundreds of large volcanoes, are thought to have been important in allowing the planet's internal heat to escape.

Inside Venus

Samples obtained by the Russian Venera landers show that Venus has a rocky crust made of basalts similar to those found under the oceans on Earth. Venus probably also has a thick mantle and a molten metallic core. Most

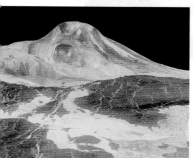

planetary scientists believe the interiors of Venus and Earth are made of similar materials.

Yet the planet has an extremely weak magnetic field. Part of the reason for this lack may be the slowness of the planet's rotation. It rotates once every 243 Earth days. Remarkably, Venus differs from the other planets in having a retrograde rotation: it spins from east to west. This rotation may be the result of a large impact early in its history.

The Venusian Atmosphere

The clouds responsible for the brilliance of Venus totally obscure its surface. Even modern telescopes show only faint, irregular cloud patterns and can never glimpse the surface below.

The Venusian atmosphere is 96.5 per cent carbon dioxide. There are three principal cloud decks, lying approximately 48–60 km (30–37 miles) above the surface, much higher than clouds on Earth. The clouds revolve around Venus at high speeds, carried by winds up to 350 km/h (220 mi/h), driven by energy from the Sun. The clouds are made from tiny droplets of sulphuric acid, making them as corrosive as an acid bath.

Although the temperature at the top of the atmosphere measures a chilly -45°C (-49°F), carbon dioxide is a potent greenhouse gas and the temperature rises steadily towards the surface. On the ground, temperatures soar to over 460°C (860°F), nearly twice as hot as a conventional oven and hot enough to melt lead. Furthermore, the atmospheric pressure at the surface of the planet reaches a crushing 90 times that experienced at sea level on Earth. Far from being Earth-like, the surface of Venus is one of the most hostile known.

EARTH

Largest of the terrestrial (rocky) planets, lying third from the Sun. Earth is a dynamic planet, with active volcanoes and areas of mountain-building. Beneath the thin crust lies the rocky mantle and an iron core. Earth's location in the inner Solar System means it has the ideal temperature for large oceans of liquid water to exist. These cover two-thirds of its surface, and a moist atmosphere supports its delicate ecosystems. It is the only place in the Solar System where life is known to exist.

Interior and Surface

By studying shock waves (seismic waves) from earthquakes, scientists can create a picture of the interior of our planet. Recording the velocities of earthquake waves, and noting positions where their velocity changes, has revealed distinct layers and boundaries in Earth's interior.

Earth's core can be divided into two parts: a solid inner core of iron and an outer core of molten iron. The radius of the entire core is about 3,500 km (2,200 miles). Above the core sits the mantle, 2,900 km (1,800 miles) thick, and consisting of iron- and magnesium-containing rocks. The mantle is not completely fluid; it acts more like putty, in that it can easily be deformed if a pressure is applied.

On top of the mantle is the crust, made of less dense granitic and basaltic rocks. The crust is divided into rigid plates that ride on the mantle. The thickness of the crust varies from about 15 km (9 miles) under the oceans to as much as 40 km (25 miles) under the continents. The overall structure of the Earth was determined early in its history. When the planet was still molten, the least dense materials rose to the surface, forming the crust, while the densest materials sank towards the centre, forming its core.

Earth's Atmosphere

Life on Earth would not be possible without the blanket of air that provides warmth, protection and the oxygen we need to breathe. The atmosphere consists of 77 per

cent nitrogen, 21 per cent oxygen, and 2 per cent other gases, including water vapour. Air pressure and density decrease with height. Very high up, the atmosphere becomes so rarefied that gas particles collide less often and are able to escape into space.

The atmosphere has several distinct layers. From Earth's surface to a height of approximately 12 km (7 miles) is the troposphere, which comprises 75 per cent of atmospheric gas. Weather occurs here and the temperature drops with increasing altitude. The stratosphere continues to 50 km (31 miles) and contains the ozone layer, which absorbs most of the harmful ultraviolet radiation from the Sun.

ABOVE: The enchanting sight of Earth from space.
BELOW: Earth's atmospheric layers. The temperature of Earth's atmosphere varies greatly with increasing altitude. These changes are shown by the curved orange line as it passes through the principal layers of the atmosphere.

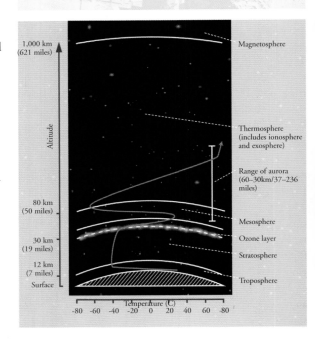

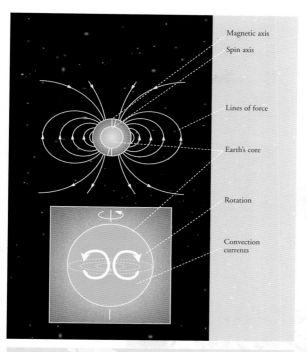

Magnetic axis

Spin axis

Lines of force

Earth's core

Rotation

Convection currents

ABOVE: Earth's rotation combines with convection currents to drive fluid motions in the molton core. These motions generate the Earth's magnetic field, whose lines of force extend beyond the surface out into space. The magnetic axis is tilted around 12° from the spin axis of the planet.

The mesosphere, from 50 to 80 km (31 to 50 miles), is the coldest layer. Up to about 1,000 km (600 miles) is the thermosphere. This layer includes both the ionosphere, which reflects radio waves, and the exosphere. Beyond this lies the magnetosphere, a magnetic bubble around the Earth that traps solar particles.

Atmospheric Circulation

Earth's rotation and heat from sunlight combine forces to affect atmospheric circulation. Through the process of convection, heated air rises while cooler, denser air sinks to replace it. Locally, convection in the atmosphere produces fluffy cumulus clouds and stormy thunderheads. On a global scale, the sunlight that falls nearly vertically on equatorial regions creates hot air, while farther north and south the air is heated less strongly by the slanting sunlight and so is colder and denser. Consequently an atmospheric flow between the equator and the poles is generated.

Oceans, Climate and Weather

Earth's oceans may be thought of as huge storage heaters, which regulate Earth's climate and day-to-day weather. Because the sea takes time to warm up and cool off, it has a warming effect during the winter, and a cooling effect during the summer. Large continental landmasses experience a greater diversity of temperatures than a small landmass surrounded by the sea.

Blue Skies, Red Sunsets

Sunlight entering Earth's atmosphere is scattered by air molecules. Since the shorter-wavelength blue light is scattered more than the longer-wavelength red light, the sky appears blue. At sunset and sunrise, when the Sun is near the horizon, the light from the Sun must pass through more atmosphere. In doing so, blue light undergoes far more scattering, so that less reaches the ground. However, red light is relatively unaffected and this results in crimson sunsets and sunrises.

Earth's Magnetic Field

Earth's magnetic field can be compared to that of an immense bar magnet. The north and south magnetic poles lie deep within the Earth, 12° askew from the corresponding north and south spin axes. Magnetic field lines, or lines of force, flow out from the south magnetic pole and loop inwards at the north magnetic pole. The field is thought to arise from currents in the fluid outer core. The field exerts a force on any electrically charged particle moving through it. Particles are redirected from their original paths and spiral along the field lines. The region of space occupied by the Earth's magnetic field is called the magnetosphere.

Measurements in rock show that Earth's poles have swapped round 171 times in the last 71 million years. During a reversal, the magnetic field's strength drops to zero.

⫸ *Plate Tectonics, p80; Continental Drift, p80; Ozone Layer, p80; Van Allen Belts, p81; Aurora, p81; Coriolis Effect, p105; Greenhouse Effect, p106; Dynamo Model, p108; Magnetosphere, p108*

AGE OF THE EARTH

Measurements from the radioactive decay of elements put the formation of the oldest rocks on Earth at just under four billion years ago. The dates have been obtained from metamorphic rocks, those formed by the baking of precursor rocks under very high temperatures and pressures, deep within Earth. From the dating of other Solar System material, such as meteorites and the Moon, the modern consensus is that the Solar System as a whole formed about 4.55 billion years ago. Earth is likely to have started forming at this time as well. Current theories suggest that it continued to grow through the bombardment of planetesimals for some 120 to 150 million years. At that time, 4.44 to 4.41 billion years ago, Earth's atmosphere began to develop and its core began to form. The rocks that formed at this stage are likely to be those that were later baked to form the oldest metamorphic rocks known today.

PLATE TECTONICS

The processes by which Earth's crust is shaped, involving interactions between the large slabs called plates into which the crust is divided. The plates are supported by the denser mantle beneath. Plates can slide past each other, creating a fault zone such as the San Andreas fault line, where the Pacific plate slides past the North American plate. The release of pent-up energy as two plates grind past each other gives rise to earthquakes. Alternatively one plate may descend under another, creating a subduction zone. A subduction zone is also known as a destructive plate margin, since the plate which is sinking is destroyed in the process. Such plate boundaries are typically marked by strings of active volcanoes, mountains and earthquake activity.

When two plates converge with little or no subduction, significant mountain building often occurs, as one plate buckles against the other. The greatest mountain belt on Earth, the impressive Himalayas, is the result of India crunching into Asia.

At a constructive plate margin, crust is formed rather than destroyed. The crust beneath the oceans is formed by the eruption of lava along mid-oceanic ridges, such as the Mid-Atlantic Ridge. The new crust is carried away from the mid-oceanic ridges towards the continents, where it will eventually be subducted and melted.

CONTINENTAL DRIFT

The slow movement of the continents over the surface of Earth. The process of plate tectonics is responsible for continental drift. The decisive evidence for continental drift comes from magnetic studies of ancient rocks. When rocks are formed, iron-rich minerals in them become magnetized in the direction of the Earth's magnetic field. Once the rocks have formed, the direction of the magnetic field is frozen into them. Thus the rocks carry an indication of their position and orientation of the Earth's magnetic field at that time. Since Earth's magnetic field also reverses periodically, the time of a rock's formation can be revealed by its magnetism.

Putting together rock-magnetism studies from all over the globe reveals the motions of the continents. They show that land-masses that exist today once formed two large continents: Gondwanaland in the southern hemisphere, and Laurasia in the northern. Before this these two formed a single mass, named Pangaea.

))))▶ *Plate Tectonics, above*

OZONE LAYER

A region of the atmosphere about 30 km (19 miles) above the surface of Earth, in which the greatest concentration of the gas ozone occurs. Potentially deadly

ultraviolet radiation strikes oxygen and is absorbed in creating O_3, or ozone. The threat to human beings from ultraviolet radiation ranges from sunburn to skin cancer. Concern is growing over human activities which alter the amount of ozone in the stratosphere. Since the 1950s, measurements over Antarctica have shown an intensification of the regular decrease in ozone that occurs in the spring. Chlorofluorocarbons (CFCs), which are substances found in aerosols, refrigerants and other materials, are the leading pollutants that result in the depletion of ozone. CFCs can remain in the stratosphere for over 100 years. Ultraviolet rays break up CFCs, releasing chlorine, which in turn breaks down ozone. One chlorine atom destroys about 100,000 ozone molecules before it is rendered harmless by being combined with nitrogen dioxide.

LEFT : The extent of the hole in the ozone layer is shown here by the blue area – this image shows the continent of Antarctica and the South Pole.

ABOVE: Aurorae are caused by charged particles trapped in the polar regions of Earth's magnetic field. The charged particles energize gases in the Earth's upper atmosphere, causing them to glow like a neon sign.

VAN ALLEN BELTS

Two torus-shaped regions within the Earth's magnetosphere in which charged particles are trapped. The outer belt contains mainly electrons captured from the solar wind. The inner belt contains both protons and electrons. Within the lower belt is a radiation belt containing particles produced by interactions between the solar wind and cosmic rays.
)))▶ *Aurora, below; Solar Wind, p70; Magnetosphere, p108*

AURORA

The solar wind is a stream of charged particles that flow away from the sun. Those paricles that encounter Earth become trapped within two regions beyond the atmosphere called the Van Allen radiation belts. Sometimes the Sun releases an intense outburst of charged particles and the Van Allen belts 'overflow'. Particles cascade down into Earth's upper atmosphere, exciting molecules of atmospheric gases and causing them to emit visible radiation. This radiation is an auroral display. Most activity appears in the skies near the poles. In the north, the display is called the Aurora Borealis, and in the south, the Aurora Australis.

)))) *Magnetosphere, p108*

TIDES

Rise and fall of the oceans under the gravitational influence of the Moon and Sun. Two high tides and two low tides occur each day.

The gravitational attraction of the Moon is stronger on the side of Earth closest to the Moon. Here the oceans are pulled in the direction of the Moon, creating a high tide. However, a high tide also occurs on the opposite side of Earth, the side that faces away from the Moon. This bulge of water forms because Earth is very slightly pulled toward the Moon, leaving an accumulation of water behind. As Earth rotates, so its land masses pass through the two bulges of water, and experience two high and two low tides a day.

In practice the tides occur slightly ahead of the Moon and not directly in line with it. Earth rotates faster than the Moon revolves and friction drags the ocean bulges ahead of the Moon. The friction also slows Earth's rotation slightly: over millennia the day is lengthening.

The gravitational attraction of the Sun also raises tides, but its effect is weaker, because the Sun is so much farther away than the Moon. The effect of the Sun's attraction is to enhance or to weaken slightly the attraction of the Moon. At full and new Moon, the Sun, the Moon and Earth are in alignment, resulting in a tidal force stronger than average. This creates spring tides, which have a greater range between high and low tides. At the Moon's first and third quarter phases, the three bodies form a right angle. With the Moon pulling in one direction and the Sun in the other, the effect on the oceans is less dramatic, resulting in less of a tidal range. These are called neap tides.

Much smaller, but finite, tides are raised in the solid body of the Earth and the Moon. Indeed, tidal forces are experienced by any body of finite size that is placed in a non-uniform gravitational field. Close binary stars (stars

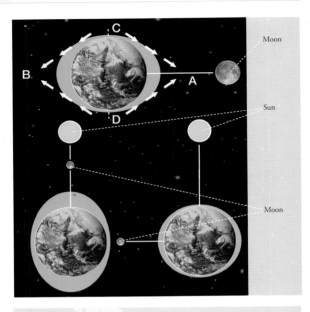

ABOVE: Tidal effects on Earth. 1. The gravitational pull of the Moon causes the oceans to bulge at points A and B, and to be reduced at points C and D. 2. When the Moon and the Sun are in line, their gravitational effects combine to produce greater tidal ranges ('spring tides'). 3. When the Moon and the Sun are at right angles, their influences partly cancel each other out, producing smaller tidal ranges ('neap tides').

that orbit close to each other under their mutual gravitational attraction) are drawn out into egg shapes by tidal interactions.

PRECESSION

Slow wobble of the Earth in space like a spinning top, under the combined gravitational pulls of the Sun and Moon. As the Earth's orientation changes, the celestial poles (which are defined by the position of the Earth's poles) trace out a complete circle against the stars every 26,000 years approximately. Not only are the positions of the celestial poles affected, the coordinates of all objects on the celestial sphere gradually change. The steady march of precession means that the positions of all stars in a catalogue, or the coordinates on a chart, have to be referred to a set date, known as the epoch.

EPHEMERIS TIME

By the 1930s it was clear that the Earth does not rotate smoothly, so that the length of the day is not constant. In the 1950s astronomers therefore introduced ephemeris time (ET), defined in terms of the predicted positions of the Sun, Moon and planets. Such predictions are listed in a table called an ephemeris. In 1984 ET was replaced for astronomical purposes by Terrestrial Time (TT) and Barycentric Dynamical Time (TDB), both of which are defined in terms of International Atomic Time (TAI), which is determined by the electromagnetic radiations given out by certain atoms.

▶ *Greenwich Mean Time, p169; Sidereal Time, p169*

EQUATOR

The great circle on the surface of a near-spherical body such as a planet or star, situated halfway between its poles. The equatorial plane is perpendicular to the body's axis of rotation and passes through the centre of the body.

TROPICS

Two circles of latitude on the surface of the Earth which correspond to the maximum northerly and southerly latitudes at which the Sun can be vertically overhead at noon at some date in the year. The Tropic of Cancer (23.5° north) is reached by the Sun at the summer solstice about 21 June, while the Tropic of

Capricorn (23.5° south) is reached at the winter solstice, about 22 December.

ECLIPTIC

The projection onto the celestial sphere of Earth's orbit around the Sun, and thus the apparent path of the Sun in the sky through the year. As the Earth's axis is inclined at an angle of about 23.5°, the ecliptic is inclined by this amount to the celestial equator. This angle is known as the obliquity of the ecliptic. The equinoxes are the points where the ecliptic crosses the celestial equator; the positions of these points change over the years because of precession.

▶ *Precession, p82*

NODE

One of the two points where an orbit intersects a reference plane – for example, the orbit of a planet with the ecliptic. The ascending node is the point at which the body moves from south to north and the descending node is where it moves from north to south.

AGE OF AQUARIUS

The point where the Sun crosses from south to north of the celestial equator each year is the spring equinox. This crossover point gradually moves against the stars, owing to precession, Earth's slow wobble in space. Over 2,000 years ago, the crossover point lay in Aries. Subsequently it moved into Pisces, where it still lies. It will eventually reach Aquarius, nearly 600 years from now – which is when the much-heralded Age of Aquarius will start.

APOGEE

The point in the orbit of the Moon or of a spacecraft that is farthest from Earth.

▶ *Perigee, below*

PERIGEE

The point in an orbit about the Earth which is the nearest to the Earth.

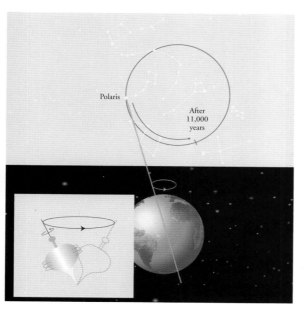

LEFT: The Precession of the Earth. Currently the North Celestial Pole lies near Polaris, the North Star. After about 11,000 years, it will be near Vega.

THE MOON

Earth's only natural satellite. With a diameter of about one-quarter of Earth, the Moon is large compared to the planet it orbits – the Earth-Moon system is often referred to as a double planet. It is the only world beyond the Earth to have been visited by human beings – in the Apollo missions (1968–1972).

Observing the Moon

With the unaided eye, two principal landforms can be seen: dark patches, which are the *maria*, and the brighter highlands, called *terrae*. Through a pair of binoculars or a small telescope, the Moon is a fascinating object to study, revealing myriads of craters and mountain ranges, as well as smooth lava flows covering the lowland plains. The best time to pick out craters and mountain ranges is when they lie close to the terminator, the line which divides day and night. The low-angled illumination helps throw the topography of the lunar surface into stark relief.

Occasionally, the Moon will pass in front of a bright star or planet. This is called an occultation. Because the Moon has a negligible atmosphere, stars passing behind it will wink out instantaneously when they encounter the limb of the Moon, while planets take longer to disappear.

The Formation of the Moon

The Moon probably formed from the debris created when a Mars-sized object impacted Earth, about 50 million years after the formation of Earth. Further material was pulled onto the Moon gravitationally from around 4.5 billion years ago. Leftover debris continued to impact the Moon, heating and melting the newly formed crust. Slowly, the impacts ebbed and the crust had a chance to cool and solidify, approximately 4.3 billion years ago. Debris that impacted the surface after this time formed the many craters seen today on the surface. The interior was still hot and molten, and some of the larger impacts fractured the lunar crust and allowed magma to flow outwards, creating the dark *maria* and lowland basins. The vulcanism which formed the *maria* lasted from 3.9 billion years ago to 3.0 billion years ago, and ended when the Moon's interior cooled so much that the crust became thick and impermeable, trapping the magma permanently below the surface.

The Surface of the Moon

The Moon's surface consists of mountainous highlands (called *terrae*) and smoother lowlands (called *maria*). The maria are lowlands flooded with basaltic lava, and many of them formed when the lunar surface was fractured by large asteroid impacts relatively early in the Moon's history.

Craters are most common on the lunar highlands, resulting from the relentless bombardment of space debris billions of years ago. There are ghostly remnant craters and youthful ray craters,

ABOVE: Although the Moon has no water on its surface, evidence suggests that water ice may exist in the deep craters at the lunar poles (centre, in blue).
ABOVE CENTRE: A close-up of the Moon in its crescent phase. Being so close to us, the Moon is ideal as a first target for the amateur astronomer – a pair of binoculars will reveal its most prominent features.

RIGHT: Samples of rock and soil collected by the astronauts who landed on the lunar surface yielded vital information about the Moon's composition.

which retain bright streaks radiating outwards.

Craters are termed 'complex' if they have flat floors, terraced walls and central peaks – huge mountains of material thrown upward during the last stages of the crater's formation. These are typically tens of kilometres to a few hundred kilometres in diameter. Two beautiful examples of complex craters are Tycho and Copernicus, each about 90 km (55 miles) in diameter. Tycho has a distinctive pattern of bright rays composed of debris sprayed thousands of kilometres across the lunar surface. Rays are typical of young craters. The ejecta has not had time to be darkened by the action of microscopic dust and cosmic rays which rain down onto the lunar surface.

Craters larger than about 320 km (200 miles) across are called impact basins and are the largest impact structures on the Moon. Most are over 800 km (500 miles) in diameter and result from truly cataclysmic collisions. They are typically flooded by smooth basaltic lava that either welled up through fractures in the lunar crust, or formed from melting of the crust when the asteroid-sized impactor struck.

Surface Composition

The surface of the Moon is covered with an ancient soil called the lunar regolith. The regolith consists of rock fragments and finer particles which have been ground down through billions of years of bombardment by micrometeorites. Round, glassy spherules are also common within the regolith. These drops are the remains of impacts that melted the surface rocks, and materials flung from volcanoes when the Moon was still geologically active.

Moon rocks can be divided into three main categories: anorthosites, basalts and breccias. The anorthosites are common in the hilly regions and are more numerous than the other types; they are pale, igneous rocks containing aluminium and calcium. Basalts are dark, dense rocks containing iron, titanium and magnesium. The basalts formed when lavas erupted on to the lunar surface and cooled between 3.9 and 3.0 billion years ago. These rocks

form the maria. Breccias are rocks reconstituted from fragments of anorthosites and basalts, shattered during violent impacts and cemented together by later impacts.

Rocks on the Moon contain no water. However, the Lunar Prospector probe found tentative evidence for water-ice mixed in with the lunar regolith at the edges of steep crater walls at the poles. If there is water there, it is likely to have come from comet impacts.

How Earth and the Moon affect each other

As with many satellites in the Solar System, the Moon is locked into a rotation pattern in which its orbital period is the same as the time it takes to turn once on its axis. This relationship is called 'synchronous' or 'captured rotation'. Synchronous rotation explains why we always see the same side of the Moon from Earth.

The same gravitational tug-of-war is also slowing the rotation of Earth. Eventually the length of the day on Earth will match the length of the month. When this happens the Moon will appear to hang permanently in the same part of the sky.

Another result of this orbital dance is that the Moon is very slowly moving away from the Earth. A small amount of gravitational energy is transferred from Earth to the Moon, causing it to recede from Earth by almost 4 cm (1.5 in) a year.

)))➤ *Tides, p82; Phases of the Moon, p86; Far Side of the Moon, p86*

PHASES OF THE MOON

The changing appearances presented by the Moon throughout the month, to do with the relative position of the Earth, Sun and Moon. The Moon shines by reflected sunlight, and as it moves in its orbit, a varying amount of its illuminated side is visible from Earth.

New Moon occurs when the Moon lies between the Sun and Earth and hence its unlit side faces us. During the next seven days, the Moon is said to be waxing, or growing larger, and the phase is called a waxing crescent. A half Moon occurs seven days after new Moon. During the ensuing week, seven to 14 days after new Moon, more of the illuminated side becomes visible. A three-quarters illuminated Moon is called a gibbous Moon. Fourteen days after new Moon, the Moon lies on the opposite side

ABOVE: The different 'phases', or shapes of the Moon result simply from the fraction of the sunlit half of the Moon that we see as it orbits the Earth.

of Earth from the Sun. The phase seen is the full Moon. Over the next two weeks the amount illuminated as seen from Earth diminishes: the Moon is said to be waning.

FAR SIDE OF THE MOON

The Moon always keeps the same face towards Earth hence from Earth we can never see the far side of the Moon. The far side is miscalled the 'dark' side of the Moon – all areas of the Moon experience two-week days followed by two-week nights. The large, dark maria – plains of solidified igneous rock – which are common on the near side are lacking on the far side. The crust there is thicker, and it was more difficult for magma to escape to the surface in the past. The far side is a nearly continuous stretch of craters of all sizes.

GIBBOUS MOON

Describing the phase of the Moon, planet or satellite when more than half, but less than the whole of its disk is illuminated as seen from Earth.

MARS

Like Earth, Mars has seasons. Its axis is tilted with respect to its orbit at almost the same angle as Earth's but, owing to the length of the Martian year, its seasons last twice as long. As Mars's poles are alternately illuminated by the Sun, the bright polar caps can be seen to wax and wane.

The Great Martian Volcanoes

Four huge volcanoes dominate Mars and are the largest in the Solar System. Arsia Mons, Pavonis Mons, Ascraeus Mons and Olympus Mons, the largest of them all, dwarf any volcano on Earth.

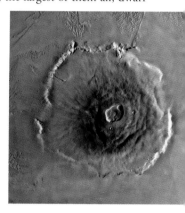

The summit of Olympus reaches up more than 27 km (17 miles) above mean surface level, and its base is 550–600 km (340–370 miles) across. In comparison the largest volcano on Earth, Hawaii's Mauna Loa, measures 120 km (75 miles) across its base and 9 km (6 miles) in height.

The fourth planet from the Sun, with a striking red colour that led the ancients to name it after the Roman god of war. One-and-a-half times farther from the Sun than Earth, Mars takes 687 days to complete one orbit. Mars is a rocky planet with only about half the Earth's diameter and one tenth of its mass. It is colder than Earth, being farther from the Sun and wrapped in a much thinner atmosphere. There are enormous volcanoes and canyons, and there is strong evidence that liquid water once flowed across its now barren surface. The existence of water suggests that there may be life on Mars.

Seeing Mars

At its closest approach to Earth, Mars can be only 56 million km (35 million miles) away, and brighter than any star. Earth, Mars and the Sun line up once every 780 days on average. Through a moderate amateur telescope, it is possible to see darker patches on the surface, and the bright polar caps.

Olympus Mons and the other giant volcanoes are similar in nature to the shield volcanoes of the Hawaiian Island chain, formed by magma welling up from a 'hotspot' formed by a rising plume of material beneath the crust. On Mars, it seems that there has been little plate tectonics – the crust has never been in horizontal movement – so a volcano above a hotspot would stay there, and keep on growing to the staggering proportions we see today.

Not all Martian volcanoes are the same. Alba Patera, north of the Tharsis bulge, is some 1,500 km (930 miles) across but shows very little vertical relief. The large number of craters peppering it suggest that its activity was at a maximum around 1.7 billion years ago. Around the volcano can be seen solidified lava flows and collapsed lava tubes.

Other volcanoes, such as Tyrrhena Patera, show evidence that their eruptions were explosive in nature. It

ABOVE: The Hubble Space Telescope has regularly monitored Mars and its atmosphere. Dust storms, cyclones and weather fronts are common throughout the Martian year.
RIGHT: Orbiting space probes have revealed volcanoes such as Olympus Mons, which may have been active as recently as 10 million years ago.

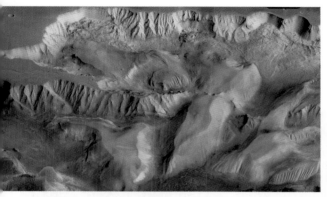

LEFT: This close-up image shows part of the vast Valles Marineris which stretches for 4,000 km (2,500 miles).
BELOW: Evidence that water once flowed on Mars comes from networks of channels on the planet's surface that resemble dried-up river beds.

is likely that explosive volcanic activity on Mars was driven by rising magma interacting with water or ice in the crust, creating a volatile mixture that erupted furiously at the surface.

Although Mars shows little signs of Earth-like plate tectonics, geological faulting has played a major part in shaping the planet's surface. Rifts and faults can be seen around many volcanoes, but the most imposing such feature is the Valles Marineris system. This is an enormous system of canyons, 4,500 km (2,800 miles) long, that would stretch from coast to coast of the USA. It is up to 7 km (4.3 miles) deep and, in places, is 600 km (370 miles) wide.

Signs of Water

Chunnels are distributed across the planet showing a variation in shape and size. Many of these channels are very similar to river valley networks on Earth. They show tributaries and increase in size downstream. But, in keeping with Mars' large-scale geology, there are other features that dwarf any terrestrial counterparts. For example, there are channels closely resembling landscapes formed by huge floods on Earth, only much larger. The presence of chaotic, jumbled terrain at the upper reaches of these Martian channels strongly suggests that the water erupted from under the ground.

The Interior

Mars' dense, iron-rich core has a diameter of around 2,900 km (1,800 miles). This is surrounded by a mantle 3,500 km (2,200 miles) thick, and in turn by a thin, light rocky crust whose average thickness is roughly 100 km

(60 miles). A very weak magnetic field, around 1/800th the strength of Earth's, is most likely the remnant of a strong field that Mars once had, before its core cooled and solidified.

Atmosphere and Weather

The atmosphere on Mars has a pressure at the surface less than 1 per cent of sea-level pressure on Earth. It is composed of more than 95 per cent carbon dioxide, with the rest consisting mainly of nitrogen and argon. Water vapour, oxygen and carbon monoxide are also present.

Thin clouds of water-ice crystals form high in the atmosphere, often capping the summit of Olympus Mons.

Dust devils frequently criss-cross the surface. But the most spectacular phenomena which will confront any future Martian meteorologist are the great dust storms, which can engulf the surface of the whole planet.

➤ *Life On Mars, p89; Martian Polar Caps, p89*

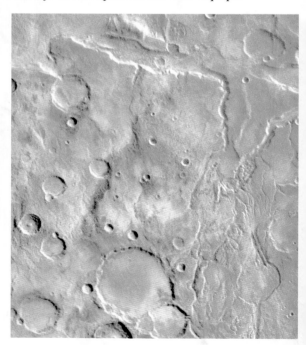

LIFE ON MARS

In July 1976 Viking 1 reached the surface of Mars carrying a mechanical scoop, with which it collected Martian soil to be analyzed in a number of on-board experiments. The initial results showed oxygen was produced by the soil, as if microbes were digesting the nutrient liquid provided in the experiment. But this was a false alarm, triggered by a simple chemical reaction. Since the Viking missions, the quest to understand life on Mars has focused on Earth life in extreme environments. In Antarctica living organisms have even been discovered inside rocks, forming a green layer just below the rock's surface. If life could adapt to these conditions, then why not to Martian environments? In 1996 NASA scientists announced that in a meteorite from Mars they had discovered microscopic structures that looked similar to Earth bacteria. However, scientists now believe that these fossil microbes contaminated the rock after it landed on Earth 13,000 years ago.

MARTIAN POLAR CAPS

The bright polar caps of Mars shrink and grow with the Martian seasons. When they are at their smallest, the northern cap is the larger of the two, around 600 km (373 miles) across. The southern residual cap is only around 400 km (250 miles) across and is thought to be composed mainly of carbon dioxide. Both caps have been layered and cut by canyons over millions of years, as ice and dust are deposited and stripped away over the changing seasons.

BELOW: Martian meteorite ALH84001 caused a stir in 1996 when scientists announced it contained a number of minute structures thought to be evidence for ancient Martian life. Later, other researchers disputed these claims.

PHOBOS

The larger of the two satellites of Mars. Phobos is irregular in shape, being 27 km (17 miles) along its greatest axis. In the distant past, Phobos and Deimos strayed from the asteroid belt, coming too close to the Red Planet and becoming its satellites. Phobos has a dark, dusty, heavily cratered surface, marked by a series of almost parallel grooves. Their most likely origin is the impact that produced crater Stickney, 5 km (3 miles) across, the largest on Phobos. Phobos orbits less than 6,000 km (3,700 miles) from the surface of Mars, circling the planet three times in a Martian day.

⟫⟫▶ *Deimos, below*

DEIMOS

The smaller of the two satellites of Mars. Deimos is potato-shaped and measures 15 x 12 x 10 km (9 x 7 x 6 miles). Like the other Martian moon, Phobos, Deimos almost certainly originated in the main asteroid belt, between Mars and Jupiter. In the distant past, Phobos and Deimos strayed too close to the planet and fell under its gravitational influence, becoming satellites of Mars. Deimos has a dark grey, dusty surface covered with impact craters. It revolves around Mars in 1.262 days and rotates on its axis in the same time, so that one face is always towards Mars.

THE GIANT PLANETS

Beyond the inner rocky planets and the asteroid belt lie four huge planets, separated by enormous distances. The inner planets, Mercury, Venus, Earth and Mars, all lie within 250 million km (155 million miles) of the Sun. But Jupiter, the next planet, orbits the Sun over three times farther out than Mars. Then comes the magnificent ringed planet Saturn, at nearly twice the distance of Jupiter.

Uranus, the next planet, is 19 times Earth's distance from the Sun. The next planet, Neptune, is half as far again as Uranus from the Sun. Beyond these four lies the outermost planet, Pluto, another tiny, rocky world.

The Gas Giants

The four giant planets are not at all like Earth. They are much larger and more massive. Jupiter has a diameter 11 times that of Earth and could contain every other planet and moon in the Sun's family. Saturn is smaller, with 57 per cent of the volume of Jupiter. Uranus and Neptune are very similar to each other, with about four times the diameter of Earth.

Rather than being rocky worlds, like those of the inner Solar System, these giants are composed mostly of gas, predominantly hydrogen and helium. The interiors of Jupiter and Saturn consist mainly of liquid hydrogen, but at their very centres there are thought to be relatively small rocky cores, perhaps the size of Earth. Uranus and Neptune are

BELOW: Gas giants, Saturn, and bottom from left, Jupiter, Uranus and Neptune. Jupiter and Saturn are the true giants, being 11 and 9 times wider than Earth respectively. Uranus and Neptune are around four times Earth's width.

composed mainly of water, ice, methane and ammonia, with perhaps 10–15 per cent less hydrogen and helium than their larger siblings.

The Birth of the Gas Giants

When the planets were forming, over 4.5 billion years ago, the inner part of the Solar System was too close to the Sun for substances such as water and methane to solidify as ice. Consequently, the inner planets are formed from rock and metal. Farther out, there was a boundary beyond which it became cold enough for ices to condense. The outer planets formed by accumulating these ices as well as the rock, and thus grew to the huge masses that they have today. Hydrogen gas in the solar nebula was also swept up by the powerful gravity of these growing gas giants, swelling them further.

The Moons of the Giant Planets

The giant planets of the outer Solar System are attended by more than 130 known moons, and it is likely that more await discovery. Many are small, irregularly shaped objects, probably captured asteroids. The larger moons display great variety. These worlds formed far from the Sun, in an environment rich in

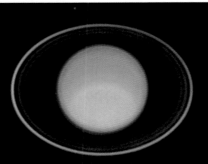

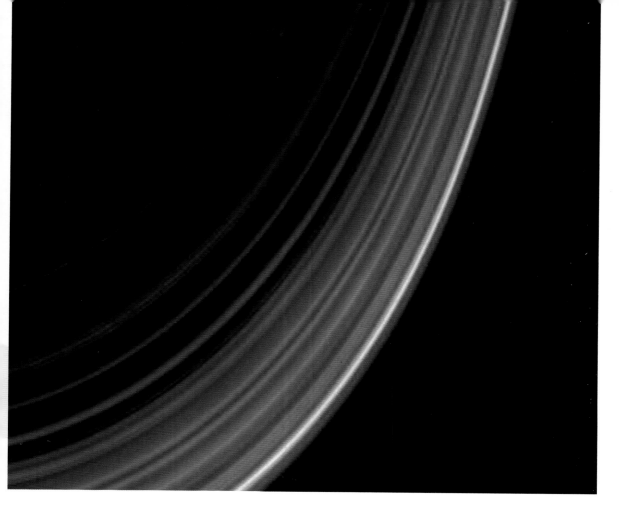

ices, and their relatively low densities suggest that they are largely made of ices.

))))➤ *Jupiter, p92; Saturn, p96; Uranus, p99; Neptune, p100*

PLANETARY RINGS

 Rings of debris, consisting of rock and ice, that orbit the four gas giants. The formation of planetary rings is still not fully understood. For many years it was considered that the rings formed when a moon or stray asteroid entered inside the Roche limit, a zone around a planet where the gravitational pull of the planet is strong enough to tear a body apart. The French mathematician Edouard Roche first proposed the idea over a century ago. For Saturn, the edge of the Roche limit does indeed lie close to the outer edge of the main ring system. However, a number of moons have now been found within the Roche limits of Saturn and other outer planets. Furthermore, the ring systems for some planets extend beyond their Roche limit. While the Roche limit is likely to play a significant role in the formation of planetary rings, by breaking up small moons and asteroids, ring formation is a more complex process and probably involves the 'shepherd moons' that orbit inside and outside the rings. The rings themselves are likely to be relatively young, geologically speaking, and are thought to be replenished by dust and fragments from the surfaces of small moons, as well as by cometary material.

))))➤ *Saturn, p96; Roche Limit, p143*

ABOVE: Saturn's rings are made from chunks of rock and ice, some as large as cities and some the size of smoke particles.

JUPITER

The fifth planet from the Sun and by far the largest in the Solar System. It is twice as massive as all the other planets combined, and so voluminous that 1,300 Earths would fit inside it. Made mainly of hydrogen and helium, Jupiter has no solid surface. Its turbulent atmosphere extends downwards until extreme temperatures and pressures form a liquid core of metallic hydrogen, inside which there may be a rocky core. Jupiter orbits the Sun once every 11.9 years, accompanied by a large entourage of satellites.

Jupiter's Atmosphere

Jupiter's rapid rotation speed – 45,000 km/h (28,000 mi/h) at the equator – flattens the planet at the poles and creates a bulging equator. It also organizes the outer atmosphere into regions of alternating wind jets, decreasing in strength towards the poles and creating a banded cloud structure. Eastward winds reach 400 km/h (250 mi/h) in the equatorial region, while at 17° latitude westward winds blow at 100 km/h (60 mi/h).

The Magnetosphere

Jupiter has the largest magnetosphere of all the planets. It stretches for some 700 million km (435 million miles) on the side away from the Sun. Within the magnetosphere are radiation belts similar to Earth's Van Allen belts, but far more intense. As with the other gas giants, the origin of Jupiter's mighty magnetosphere lies within its liquid interior, where a continuous churning driven by internal heat produces a dynamo effect, creating the magnetic field.

ABOVE: Despite its massive size (1,300 times the size of Earth), Jupiter is simply an enormous globe of gas, mainly hydrogen and helium.
RIGHT: Each of Jupiter's four Galilean satellites has its own distinctive features. The surface of Io pictured here is sculpted by volcanic activity.

Jupiter's Interior

In December 1995 the Galileo orbiter released a small atmospheric probe that made a one-way trip into the clouds. The Orbiter sent back a wealth of data for 57 minutes before being crushed and then vaporized by the ever-increasing temperature and pressure, about 200 km (125 miles) beneath the visible surface.

The Jovian atmosphere comprises three principal cloud decks. The uppermost of these decks is composed of cold, wispy cirrus clouds of ammonia ice. In an atmosphere whose initial mixture of elements was similar to that of the Sun, hydrogen has combined with nitrogen and carbon to form ammonia and methane. At -150°C (-238°F), the ammonia condenses and forms Jupiter's opaque upper cloud deck.

The middle cloud deck is made of ammonium hydrosulphide, formed by water vapour combining with sulphur. The pressure is about 1 bar (1 bar is the atmospheric pressure at sea level on Earth).

In the lowest cloud deck temperatures hover close to the freezing point of water. The clouds are composed of

water in the form of ice crystals or possibly liquid droplets. Farther down, the temperature, pressure and wind speed increase rapidly as Jupiter's immense gravity crushes the atmospheric gases. The Galileo atmospheric probe's dying signal recorded a temperature of 300°C (572°F) and a pressure of 22 bars, 150 km (93 miles) below the height at which the probe began taking data. Wind speeds had increased from 360 km/h (224 mi/h) near the top of the atmosphere to 540 km/h (336 mi/h) and powerful lightning strikes were recorded.

Only theory can tell us what lies deeper within Jupiter. The models suggest that a huge hydrogen- and helium-rich envelope exists, thousands of kilometres deep, becoming increasingly compressed with depth.

Eventually, the atmosphere gives way to a global sea of hydrogen, crushed to such a degree that it is liquid even though it is at temperatures of many thousands of degrees. Deeper down, hydrogen behaves like a liquid metal (liquid metallic hydrogen) which conducts electricity. Circulating flows within the liquid metallic hydrogen zone generate Jupiter's powerful magnetic field. It is estimated that the central core must be at a temperature of 24,000°C (43,000°F) and a pressure of 100 million bars.

Rings and Minor Satellites

Jupiter's collection of more than 60 moons can be subdivided into four sets. The innermost set comprises four small irregular satellites (Metis, Adrastea, Amalthea and Thebe). Then come four large satellites discovered by Galileo, there are also two outer sets of moons.

The Innermost Satellites

Metis and Adrastea, the two innermost satellites, have diameters of less than 50 km (30 miles). They are closely associated with a system of tenuous rings, so thin and dark that they were not observed until the spacecraft flybys. Among the outer satellites, orbital similarities suggest that the satellites are fragments of larger bodies. For example, they revolve around Jupiter in a retrograde direction (that is, in the opposite direction to the other satellites and opposite to the direction of Jupiter's own rotation). Satellites born together with the parent planet and other satellites would share the same direction of revolution.

Seeing Jupiter and its Satellites

Jupiter is one of the brightest objects in the sky, outshone only by the Sun, the Moon and Venus. Through a telescope Jupiter is a fine sight. Its complex cloud systems are clearly visible as light and dark bands, running parallel with the planet's equator. Perhaps the most readily distinguishable feature on the disk of Jupiter is the Great Red Spot.

Accompanying Jupiter are four easily observed satellites: Io, Europa, Ganymede and Callisto. These are known as the four Galilean satellites, named in honour of Galileo Galilei (1564–1642), who discovered them in 1610.

)))➤ *Van Allen Belts, p81; Planetary Rings, p91; Great Red Spot, p94; Galilean Satellites, p94; Magnetosphere, p108*

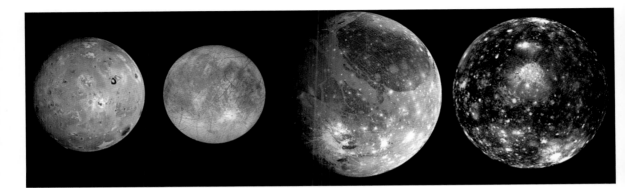

GREAT RED SPOT

Enormous storm system on Jupiter that has been observed since telescopes were first strong enough to see it, in the seventeenth century. The colour may come from trace molecules (possibly sulphur) dredged up from below. The spot extends 23,000 km (14,300 miles)

ABOVE: Galilean satellites. Galileo discovered the four major satellites orbiting the giant planet Jupiter in 1610. From the left they are Io, Europa, Ganymede and Callisto.

east–west, roughly twice the diameter of Earth, and 12,400 km (7,700 miles) north–south. There is a windspeed difference of 350 km/h (220 mi/h) across the north–south dimension of the spot. Constant replenishment from below and lack of a solid surface to disrupt the flow allow the Red Spot and other oval weather systems on the planet to exist for centuries or longer.

)))➤ *Jupiter, p92; Great Dark Spot, p101*

GALILEAN SATELLITES

The four largest satellites of the giant planet Jupiter. They are so called after Galileo Galilei (1564–1642), who observed them when he made his first telescopic observations of the sky in 1609–10. In order outwards from Jupiter they are named Io, Europa, Ganymede and Callisto, after mythological figures associated with Zeus, the Greek equivalent of Jupiter.

)))➤ *Io, Europa, Ganymede, Callisto, pp94–95*

IO

Innermost of the four large satellites of the giant planet Jupiter. Unlike the other satellites, Io's surface is a cauldron of geological activity, where active volcanic vents coat the surface in red and yellow sulphur lava. Io's slightly noncircular orbit takes the satellite through constantly changing strengths of Jupiter's gravity field and constantly flexes it. The tidal forces heat the interior and trigger volcanic activity.

ABOVE: The satellite Io. Scientists have recently discovered new details of a giant crater on its surface which contains a multi-coloured lake of molton lava.

EUROPA

One of the four largest satellites of Jupiter. Europa has a cracked, icy surface with very few impact craters, evidence of reworking of the satellite's entire surface by geological processes. Recent data from the Galileo spacecraft suggest a water ocean may exist below the frozen surface, possibly harbouring life.

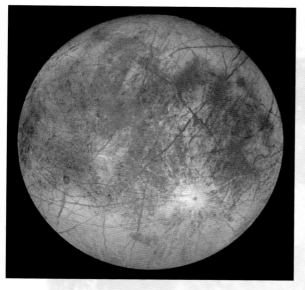

ABOVE: Jupiter's satellite Europa has a fractured and icy crust. Evidence suggests that beneath the icy surface there may be an ocean of liquid water.

GANYMEDE

One of the four large satellites of the giant planet Jupiter. It has two distinct surface components: areas with abundant craters, and others which are far less cratered, where melting and refreezing have modified the surface. The satellite's diameter is approximately 5,262 km (3,270 miles); its mean distance from Jupiter is 1.07 million km (665,000 miles).

CALLISTO

Callisto is the farthest from Jupiter of the planet's four large satellites. It therefore experiences the least tidal action and hence has the lowest internal heating. It has a dark, cratered surface, indicating that

ABOVE: The Galilean satellite Ganymede.

little geological activity has occurred there for billions of years. Its diameter is approximately 4,800 km (2,986 miles); its mean distance from Jupiter is 1.88 million km (1.17 million miles).

)))⟩ *Tides, p82*

ABOVE: The dark, cratered surface of Callisto, one of Jupiter's four largest moons, also known as the Galilean moons.

SATURN

The outermost planet that is easily visible with the naked eye, and the second largest in the Solar System. It is in many ways a smaller version of Jupiter, composed principally of hydrogen and helium and with a similar internal structure. While not unique to Saturn, its magnificent ring system is by far the most extensive, complex and brightest of all. Saturn boasts a large family of moons, including Titan, the only satellite in the Solar System to possess a thick atmosphere.

Seeing Saturn

Saturn is readily visible in the night sky. When viewed with a small telescope, it is a spectacular object. Faint bands of colour similar to those of Jupiter, but more subdued, can be seen crossing its yellow disk. The most striking feature is the planet's ring system, which, with a moderate-size telescope, may be resolved into several distinct parts. Titan, Saturn's largest moon, can also be seen as a bright point of light.

Images of Saturn show that the planet is flattened. This is because Saturn rotates on its axis in just over 10 hours, which results in an equatorial bulge: Saturn's equatorial diameter is nearly 10 per cent larger than the polar diameter.

Saturn's Atmosphere

Like Jupiter, Saturn has three main decks of cloud. However, Saturn's frigid environment causes water and ammonia clouds to form lower down than within Jupiter's atmosphere. Convective motion, fuelled by the outward transport of heat, is not strong enough to produce towering cumulus clouds in the Saturnian stratosphere. Individual clouds are far less common than on Jupiter and when they do occur they are generally short-lived. Many change rapidly in response to prevailing winds.

Measurements of the motion of the clouds reveal a broad equatorial region with eastward winds as high as 1,800 km/h (1,100 mph). These winds form alternating east-west jets, with wind speeds decreasing towards the poles. There are oval storms near 70° north latitude and 45° to 55° latitude, forming cloud systems similar to those on Jupiter. However, they are smaller, with diameters less than 5,000 km (3,000 miles).

Saturn's Interior

Although Saturn's mass is 95 times that of Earth, its average density is less than that of water. If it were possible to place Saturn on an ocean large enough, it would float. The low density and variation in gravity sensed by passing spacecraft indicate that Saturn, like Jupiter, has a small dense core surrounded by a compressed gaseous envelope which is rich in hydrogen and helium.

Saturn radiates 1.8 times more energy than it absorbs from the Sun. Decay of radioactive isotopes or slow overall contraction of the planet under its own gravity could account for the excess energy. In addition, there may be a region deep in the interior where convective mixing is so small that the heavier helium sinks toward the core, releasing energy. At shallower depths, where the magnetic field is generated, there is evidence of organized convection. Unlike Earth, Saturn's magnetic field is aligned along its rotational axis.

Storms on Saturn

In 1990, a white cloud in Saturn's equatorial region was discovered and its growth noted by amateur observers. Cameras on board the newly launched Hubble Space Telescope were used to study the development of the storm. Although ground-based observers had reported occasional storms, this was the largest storm in 57 years.

Most storms on Saturn are short-lived and disperse in a manner that suggests they are generated by convection, which transports heat from the interior. The convective activity carries material to high altitudes, where the temperature is so low that ammonia and water immediately freeze to form ice. As the rising mass encounters the prevailing winds, the ice clouds serve as markers to reveal the progress of the storm. As the 1990 storm developed, it became apparent that it was similar to two previous equatorial disturbances that had been observed within the previous 125 years. These three storms were spaced at intervals of 57 years, nearly two Saturnian years, implying that these storms may be cyclic in nature.

The Nature of the Rings

In 1610 Galileo observed Saturn with his low-power telescope and saw two protrusions, one at each side of the planet. In 1612 the protrusions had vanished. Continuing observations revealed that their apparent size waxed and waned over a period of about 15 years. Finally, in 1659,

ABOVE: A false-colour image of Saturn, taken by Voyager 2 in August 1981 from a range of 114.7 million km (9.1 million miles). Seen hovering by the planet are two of its satellites, Dione (above) and Encladus (below).

Christiaan Huygens (1629–95), a Dutch astronomer, realized that there is a thin ring around Saturn's equatorial plane, which seemed to be larger or smaller according to its angle with the line of sight from Earth.

Coinciding with these discoveries, Johannes Kepler (1571–1630) had formulated his laws of planetary motion. Although the dimensions of Saturn's ring were not known, it was apparent that the distance around the outer perimeter of the main ring system was 1.5 times greater than that around the inner edge. Kepler's laws required the inner part of the ring to move faster than the outer part. Thus, the ring could not be a rigid sheet but must be composed of a swarm of particles revolving about the planet. Observational proof of this was not obtained until 1895, when James Keeler (1857–1900) used a spectrograph to show that the orbital speed of the particles decreased outwards across the rings.

)))➧ *Planetary Rings, p91; Saturn's Rings, p98; Uranus's Rings, p99*

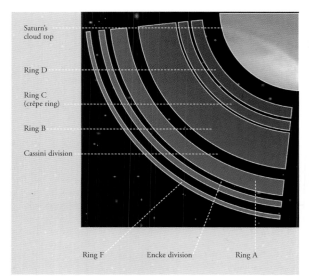

Saturn's
cloud top

Ring D

Ring C
(crêpe ring)

Ring B

Cassini division

Ring F   Encke division   Ring A

LEFT: Ring A is the outermost part of Saturn's ring system visible from Earth. The Cassini division, about 4,500 km (2,800 miles) wide, separates this from the brightest and widest part, ring B. The next ring is C, or Crêpe ring. The fainter rings D and F lie inside and outside the visible rings.

rings A and B. It was discovered in 1675 by G. D. Cassini (1625–1712).

TITAN

Saturn's largest satellite. Its diameter is one and a half times that of the Moon. Titan revolves around Saturn at a distance of over a million km (625,000 miles). It is tidally locked and rotates on its axis in 16 days, in which it revolves around Saturn. Titan has retained a thick, nitrogen-rich atmosphere that exerts a surface pressure approximately 50 per cent greater than that of Earth's. Solar ultraviolet radiation has interacted with gases in the upper atmosphere, forming thick smog. Titan's atmosphere is believed to be similar to that of Earth's shortly after its formation. The ESA's Huygens probe landed there in 2005.

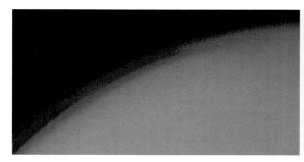

ABOVE: The surface of Saturn's largest satellite, Titan, is obscured by a haze in the atmosphere.

SATURN'S RINGS

Saturn's ring system extends from about 7,000 km (4,000 miles) above the cloud tops out to 74,000 km (46,250 miles). They are made of fragments of dark ice and rock, ranging from bodies a few kilometres across to dust particles. The rings are in the plane of Saturn's equator, which is tilted by nearly 27° compared with the planet's orbital plane.

Three distinct rings are visible from Earth: A, B and C, in order from the outermost to the innermost. A fainter inner ring, the D ring, was discovered in 1969. Farther out beyond the A ring, in order of increasing distance, are rings F, G and E, discovered by space probes. The thin, hazy E ring extends to more than 420,000 km (261,000 miles) from Saturn's cloud tops. There are also gaps, filled with thousands of thin individual 'ringlets'. Braided and kinked rings, and strange, ephemeral spoke patterns are found. The F ring has gaps and regions that appear braided or twisted. The culprits are tiny 'shepherd moons', orbiting just inside and outside the F ring. The spokes may be electrically charged dust levitating above the rings.

CASSINI DIVISION

The rings surrounding the planet Saturn consist of three main subrings, called the A, B and C rings. Earth-based observations show a number of gaps within the rings, notably the Cassini division between

HYPERION

A small satellite of Saturn with a maximum diameter of about 370 km. Its main interest is that it follows a markedly elliptical orbit, under the influence of the large moon Titan, instead of rotating smoothly around a single axis. Its irregular shape and chaotic motion suggest that Hyperion was once a larger body that suffered a number of catastrophic collisions.

➽ *Kepler's Laws of Planetary Motion, p17*

URANUS

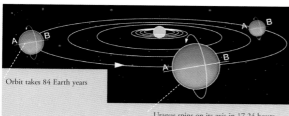

Orbit takes 84 Earth years

Uranus spins on its axis in 17.24 hours

A blue-green gas giant, the seventh planet from the Sun. It is considerably smaller than Jupiter or Saturn, but still approximately four times the diameter of Earth. Uranus is within the range of the naked eye on a moonless night, away from areas of light pollution.

The Structure of Uranus

The composition of Uranus is similar to that of Neptune. It is composed mainly of hydrogen and helium, together with methane and ammonia. It contains less hydrogen than Jupiter or Saturn. Some scientists have speculated that beneath its hydrogen atmosphere there may be an ocean of water, with methane and ammonia, over a central rocky core.

Cloud decks form deep in the atmosphere, below the upper atmospheric methane haze. The clouds are driven by wind speeds of hundreds of kilometres per hour.

The Rotation of Uranus

Uranus is unique among the Jovian planets in that its axis of rotation is tipped by more than 90°. It is possible that Uranus was struck by a large body, which 'flipped' the planet over on its side. Owing to this extreme axial orientation of Uranus, the Sun shines directly down on its North and South Poles in turn, as the planet makes its 84-year orbit. Each pole has repeated cycles of 42 years of daylight and 42 years of darkness.

The magnetic field of Uranus is tilted by 60° with respect to its rotational axis – more than any other planet

ABOVE: Rather than spinning like a top as it orbits the Sun, Uranus has a distinctive feature; its axis of spin is tilted almost parallel to its orbital plane. This means that the polar regions (points A and B) alternately point towards and away from the Sun as Uranus orbits. During an 84-Earth-year orbit, each pole experiences one 42-year day and one 42-year night.

in the Solar System. As with Earth, Jupiter and Saturn, the solar wind distorts the magnetosphere, creating a magnetotail that stretches away from the planet for many millions of kilometres in the direction opposite to the Sun.

The Moons of Uranus

Before Voyager 2 arrived at Uranus, the planet was known to have five moons. Oberon and Titania, the outermost, are covered with impact craters and stress fractures. Umbriel and Ariel, a slightly smaller pair, have distinctly different characters. Umbriel's surface appears dark and old while Ariel displays a maze of fault lines and signs of melting and resurfacing – evidence that it has undergone tidal heating that helps keep its surface geologically young. Miranda, the smallest, innermost satellite, has a spectacular surface. Composed of seemingly unrelated structures, it appears to have been shattered by a major collision and then reconsolidated, resulting in the highly complex and disordered surface seen today. Voyager 2 discovered many smaller moons. Since then, another six moons have been discovered, bringing the planet's total complement of known moons to 27 by 2005.

⟫⟫ *Giant Planets, p90; Neptune, p100*

Uranus's Rings

In 1977 astronomers organized an effort to observe Uranus as it passed in front of a star. Such an event is known as an occultation, and from it they hoped to measure the diameter of the planet more precisely and learn more about its upper atmosphere. Surprisingly, the brightness of the star abruptly dropped and returned to normal five times before it was occulted by Uranus. When the star emerged from behind Uranus, a similar pattern of fluctuations recurred, indicating the existence of five thin rings around Uranus. Additional analysis of the data revealed four more rings. In 1989 instruments on the Voyager space probe revealed two additional, faint rings. The amounts of light blocked by the rings revealed their differing thicknesses. The width of Epsilon, the densest ring, is about 100 km (60 miles) while the fainter rings are 12 km (8 miles) wide or less. The intervening gaps are hundreds of kilometres wide.

NEPTUNE

A blue-green gas giant, the eighth planet from the Sun. Its discovery was a triumph for Newton's theory of gravitation. The planet's existence and position had been independently predicted by an English mathematician, John Couch Adams (1819–92), and a French astronomer, Urbain Le Verrier (1811–77), on the basis of its gravitational perturbation of the movements of Uranus, the seventh planet. The planet was seen through the telescope in 1846. Neptune is smaller than Jupiter or Saturn, but still approximately four times larger than Earth. Neptune is too faint to be seen without the aid of a telescope or good pair of binoculars. Even the largest telescopes only reveal a small blue-green disk and occasional vague markings. Neptune has a system of rings and a suite of 13 known moons.

The Atmosphere of Neptune

Neptune is composed mainly of hydrogen and helium, with its distinctive colour arising from small amounts of methane, which absorbs red light.

BELOW: The Voyager 2 spacecraft flew past Neptune in 1989, revealing a Great Dark Spot rimmed with white cirrus clouds.

Owing to the incredibly low temperature, most clouds on Neptune form at significantly lower altitudes than those on Jupiter and Saturn, where temperatures and pressures are greater. The lower altitude means the cloud decks are harder to spot through the upper atmospheric methane haze. Once the clouds had been discovered, scientists tracked their movement, finding wind speeds of up to 580 km/h (360 mi/h).

The Great Dark Spot

Voyager 2 encountered Neptune in August 1989. Surprisingly, Neptune's atmosphere was found to be highly active. There is a high haze of frozen methane as well as high white cirrus clouds. But the most startling discovery was a series of high-pressure storm systems in the southern hemisphere, powered by heat rising from the warmer interior. The largest was the Great Dark Spot, a large 'hurricane' which whipped around the planet in a little over 18 hours. It appeared to be a region where gases were welling up. White clouds rimming the storm system suggested that methane was being forced up from warmer depths, and then froze to create cirrus clouds of ice crystals. There were other, similar storms on the planet. Unlike Jupiter's Great Red Spot, the storm moved north and south as well as east and west. Further south was the 'Scooter', another smaller storm travelling even faster, revolving around Neptune once every 16 hours. Some of the strongest winds in the Solar System were recorded close to the Great Dark Spot, a staggering 2,000 km/h (1,250 mi/h). Although the Great Dark Spot was similar to Jupiter's Red Spot, ground-based infrared

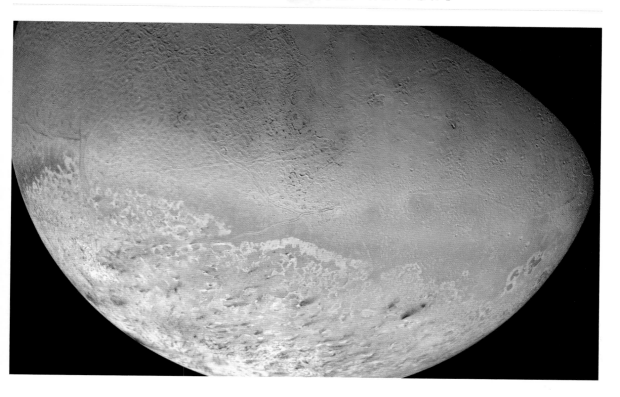

observations indicate that, like storm systems on Earth, it has dissolved. Recent images from the Hubble Space Telescope show no trace of this feature but reveal new dark spots.

The Interior of Neptune

Neptune is denser than Jupiter and Saturn. Owing to its greater density, scientists have speculated that beneath its hydrogen atmosphere an ocean of water with ammonia and methane may exist, over a central rocky core.

))))➤ *Great Red Spot, p94; Uranus, p99*

TRITON

The only large satellite of Neptune. This exotic world is about three-quarters the diameter of our Moon. Triton revolves around Neptune in a retrograde fashion – that is, in the opposite direction to the rotation of the planet. This suggests it may have been captured by Neptune in the past. Even though it currently holds the record as having the lowest observed temperature in the Solar System, -235°C (-391°F), Triton has an extremely tenuous atmosphere and some geological activity at its surface.

Of particular interest are its active geysers, which spray dark gaseous material up to 8 km (5 miles) into the atmosphere, where it encounters an airflow that sweeps the plume sideways. Fallout from the plumes can be seen as dark diffuse streaks on Triton's frozen surface. A proposed mechanism for the formation of the geysers involves transparent surface ices. Sunlight penetrates the transparent ice causing the surface beneath to sublime as the temperature increases slightly. The gas builds up under the ice and eventually the pressure ruptures the overlying ice, allowing the trapped gas to rush out, carrying fine, dark dust with it. Light is reflected from the surface of Triton in a way similar to that of Pluto and its moon Charon, indicating that these bodies may have a similar origin.

ABOVE: Triton, Neptune's largest Moon, has a surface covered with frozen nitrogen and methane, with a temperature of -235°C (-391°F).

PLUTO

 The outermost 'planet' of the Solar System, ninth from the Sun, Pluto is in fact a 'dwarf planet'. This icy world orbits the Sun once every 248 years and has the most highly inclined and eccentric orbit of all the planets. At its nearest, Pluto strays closer to the Sun than Neptune does. Its size and peculiar orbital characteristics indicate that it belongs to a group of small, icy bodies moving outside the orbit of Neptune and making up the Edgeworth-Kuiper belt. In 1978, Pluto was found to have a large moon, Charon, Pluto's only known moon until 2005, when astronomers discovered two smaller moons. Pluto is far too small and remote to be seen with the naked eye. Even large telescopes reveal it as nothing more than a faint point of light. It was discovered as late as 1930 by the US observer Clyde Tombaugh (1906–97).

BELOW: Pluto and its satellite Charon. Charon, discovered in 1978, is believed to be formed from ice thrown off Pluto when another object collided with the planet. The surface of the satellite itself is covered in water-ice.

The Atmosphere of Pluto

Pluto has a transient atmosphere. In the early 1980s methane gas was observed while the planet was nearing its closest approach to the Sun, the period of maximum warming. Because the heating power of the Sun varies by more than 60 per cent throughout Pluto's 248-year orbit, temperatures will drop considerably as Pluto moves away from the Sun and the methane gas should then condense and form fresh white ice on the surface.

Pluto is over seven times as massive as Charon, and has a brighter surface. Observations have also revealed that Charon loses methane from its surface by evaporation. Some is attracted toward Pluto, forming a frost on its surface. Nitrogen and carbon monoxide may also be present in the atmospheres of these two worlds.

)))) *Edgeworth-Kuiper Belt, p111*

The Structure of Pluto

Pluto and Charon are likely to be similar worlds in terms of surface and internal properties. The Hubble Space Telescope has been able to distinguish surface markings on Pluto, which are likely to be a combination of frost deposits and the result of collisions with smaller bodies which have modified the surface of the planet over time. An indication of what Pluto may be composed of comes from the planet's density, which is approximately twice that of water. The density indicates that Pluto is probably a mixture of rock and ice, much like Neptune's largest moon, Triton. The brighter areas distinguished by the Hubble Space Telescope are likely to be areas of frozen methane ice. Below an icy crust, Pluto and Charon may have large rocky cores.

CHARON

The largest and, until 2005, the only known satellite of the outermost planet, Pluto. Charon was discovered in 1978. The plane of revolution of the Pluto-Charon system is inclined at 99° relative to Pluto's orbit around the Sun, so every 124 years the plane lies along our line of sight, and the two bodies alternately eclipse each other. By good luck, mutual occultations began to occur soon after Charon's discovery, from 1985 to 1991. Observing the variations in the total reflected light during the eclipses gave insights into the two worlds.

Charon is 1,186 km (737 miles) across. The diameter of Pluto is less than twice that of Charon, making the system virtually a double planet. They orbit each other at an average distance of 19,600 km (12,180 miles). Each world keeps the same side facing towards the other: from Pluto, Charon always appears in the same part of the sky.

THE ORIGIN AND NATURE OF PLANETS

PANSPERMIA THEORY

The Panspermia theory claims that in the early Solar System, before the inner planets took on their present environments, life was likely to have started on all of them. The blizzards of asteroids and comets in those first few hundred million years ensured that impacts were common and oblique collisions would have blasted rocks housing simple bacteria between planets. Support for the Panspermia theory grew in the 1970s with the discovery of dormant bacteria on Surveyor 3's camera brought back after three years on the Moon. Over two decades later the discovery of bacteria in a Martian meteorite, although now discredited, kicked off the Panspermia debate again.

))))◆ *Life on Mars, p89*

PLANETESIMALS

Small bodies, composed of dust, rock or ice and ranging in diameter from less than a millimetre to several kilometres, from which the planets formed in the early Solar System by the process of accretion (accumulation of material by gravitational attraction).

))))◆ *Accretion Disk, p146*

PROTOPLANETARY DISK

A disk of dust and gas around a young star, from which planets may form. Over 100 have been observed in the Orion Nebula by the Hubble Space Telescope. They are recognizable by their strong emission at infrared wavelengths.

CRATER

Scar left when a projectile slams into a solid surface at speeds of many tens of kilometres per second. The impact causes a violent shock wave, creating a pressure millions of times greater than that of the Earth's atmosphere. The projectile either melts or vaporizes. A crater starts to form when rapid decompression of the surface rocks occurs after the initial impact. The decompression 'frees' the recently struck compressed surface, allowing a cone of rebounding debris to be flung outwards from the impact site, forming an ejecta blanket. Very little, if any, of the projectile remains. The

shape and size of a crater varies depending on the size and speed of the impactor, what it is made of (for example, ice, rock, or iron) and the type of surface it strikes (hard or soft). Icy comets are less dense than an iron asteroid and, all things being equal, inflict less damage on a planetary surface.

LATE HEAVY BOMBARDMENT

Bombardment of the bodies in the inner Solar System by debris left over from the formation of the planets which took place between the time of their formation until about 3.9 billion years ago. The oldest surfaces in the inner Solar System, those of Mars, Mercury and the Moon, bear scarring that is a record of the Late Heavy Bombardment. Between 3.9 and 3.4 billion years ago the number of impacts declined greatly. It is between these dates that the earliest known fossils are found on Earth, indicating that, as soon as the period of heavy bombardment ceased, life was able to gain a foothold.

ABOVE: Craters such as this, which is known as Ptolemaeus, are a defining feature of the lunar surface. They are formed by the numerous impacts of other bodies on the Moon over billions of years.

APHELION

The point in the orbit of a planet, comet or other celestial body that is farthest from the Sun.

))) *Perihelion, below*

PERIHELION

The point in an orbit about the Sun which is the nearest to the Sun.

CONJUNCTION

The alignment of two celestial objects (for example, of two planets or of a planet and the Sun) as seen from Earth. An inner planet – Mercury or Venus – is at 'inferior' conjunction when it lies between Earth and the Sun, and at 'superior' conjunction when it lies on the opposite side of the Sun. An outer planet cannot pass between the Earth and the Sun. Therefore, when an outer planet is at conjunction it must lie on the far side of the Sun (superior conjunction).

))) *Opposition, below; Elongation, below*

OPPOSITION

The position of a planet when it is directly opposite the Sun as seen from Earth. The outer planets – Mars, Jupiter, Saturn, Uranus and Neptune – are at their apparent brightest when they reach opposition and are best observed then. The planet then rises as the Sun sets and sets at dawn, allowing for optimum viewing and maximum viewing time.

ELONGATION

1. The angular distance between the Sun and a Solar System object (usually a planet) measured from 0° to 180° east or west of the Sun. A planet with elongation 0° is at conjunction. An inferior planet has a greatest elongation that is less than 90° east or west of the Sun. A superior planet with elongation 180° is at opposition; with elongation 90° or 270° it is at quadrature.
2. The angular distance between a planet and one of its satellites, measured from 0° east or west of the planet.

RIGHT: A total lunar eclipse can occur only when the Moon is full and Earth is lying directly between the Moon and the Sun.

LAGRANGIAN POINTS

Neutral points in the combined gravitational fields of a system of celestial bodies at which a small body will experience effectively zero net gravitational force. In a system of two massive orbiting bodies, the Lagrangian points are positions at which a smaller body can lie in equilibrium. For example, the Trojan asteroids are groups of asteroids that reside at two of the Lagrangian points in the Jupiter-Sun system, 60° ahead of and behind Jupiter in its orbit. In a system of close binary stars, the Lagrangian points are places where material can pass from one star to the other.

))) *Equipotential Surface, p117*

OCCULTATION

The obscuring of a celestial body by another. A solar eclipse is an occultation in which the Moon passes in front of the Sun as seen from Earth. Occultations also occur when the Moon, or a planet, passes in front of stars. Grazing occultations occur when a body skims the limb (edge) of another.

LUNAR ECLIPSE

A lunar eclipse occurs when the Moon passes into the shadow of Earth. This can happen only at full Moon. However, a lunar eclipse does not occur every time the Moon is full, because the Moon does not orbit in the plane of the ecliptic, the Sun's apparent path in the sky; it dips above and below it by about five degrees. So the two conditions of full phase and the Moon crossing the ecliptic

INTERPLANETARY MATTER

INTERPLANETARY DUST

Fine grains of matter that permeate the space between the planets. When interplanetary dust particles enter Earth's atmosphere at high speeds, they burn up by friction at around 100 km (60 miles) altitude, creating a trail of ionized air that glows briefly to create a meteor or 'shooting star'. Some of this dust comes from asteroids and some from comets. Each time Earth crosses a comet orbit it encounters a swarm of dust particles and we see a meteor shower.

)))➤ *Meteorites, p111; Meteor Showers, p113*

ASTEROIDS

Through a telescope, asteroids appear starlike, which gave rise to the name, which means 'little star'. They are in fact irregularly shaped and cratered chunks of rock and metal. Over 10,000 are currently known, and countless more remain to be observed. Most orbit in the asteroid belt between Mars and Jupiter. Jupiter's gravitational pull is thought to have prevented the material of the asteroids from accreting into a planet. Early in the Solar System's history there were probably fewer

ABOVE: The asteroid Gaspa, photographed by the Galileo space probe, is just one of a myriad of asteroids which orbit the sun.

asteroids than today but of larger size. Collisions have since fragmented them and many of the fragments have been swept up by the planets. The largest main-belt asteroid is Ceres, 960 km (597 miles) across. Next in order of size are Pallas, 570 km (354 miles) and Vesta, 530 km (329 miles). Numbers increase rapidly as the size decreases.

Astronomers learn what asteroids are made of by analysing the spectrum of sunlight reflected from their surfaces. Their compositions turn out to be diverse, but there are several families that show certain similarities. For example, the C-type asteroids contain carbon; S-type asteroids are mixtures of rock and metal; while M-type asteroids consist largely or wholly of metal (iron and nickel). Our most detailed information about asteroid compositions comes from pieces that have been knocked off in collisions and have fallen to Earth as meteorites.

Outside the main belt, a group of asteroids called the Trojans orbits at the same distance as Jupiter but 60° ahead of or behind it in its orbit. Among the outer planets beyond Jupiter orbits a small group of icy asteroids called the Centaurs, which may have more in common with cometary nuclei than with the solid inner asteroids. The first of these to be discovered was Chiron, in 1977, which has shown signs of comet-like activity.

)))➤ *Edgeworth-Kuiper Belt, p111*

ASTEROID BELT

The region of the Solar System between about 2 and 3.3 times the distance of Earth from the Sun in which the majority of known asteroids move. Asteroids in the belt are known as main-belt asteroids. Within the belt exist gaps (the Kirkwood gaps) resulting from the gravitational pull of the Sun and Jupiter.

KIRKWOOD GAPS

Regions within the asteroid belt where few, if any, asteroids are found. The gaps result from periodic disturbances by Jupiter, which make stable orbits impossible. They occur at distances from the Sun at which the orbital period is a simple fraction of Jupiter's orbital period: for example, $\frac{1}{4}$, $\frac{1}{3}$, $\frac{2}{5}$, $\frac{1}{2}$ etc. They are named after Daniel Kirkwood (1814–95), who first explained them.

NEAR–EARTH ASTEROIDS

Asteroids that have been thrown out of the main asteroid belt by the gravitational influences of Jupiter and Mars into orbits that can bring them closer to Earth. There are three such groups of near-Earth asteroids (NEAs): the Amor asteroids cross the orbit of Mars but not that of Earth, while the Apollo and Aten groups cross Earth's orbit. Astronomers are on the lookout for NEAs because of the collision threat they pose to Earth.

COMETS

A low-mass member of the Solar System, moving in a highly elliptical orbit and undergoing great changes in appearance as it approaches and then recedes from the Sun. Comets are highly insubstantial. Their only solid part is the nucleus, typically no more than a few kilometres

BELOW: Some comets, those moving on parabolic or hyperbolic orbits, have enough energy to escape from the Solar System.

across – a 'dirty snowball' of ice with a dusty crust.

Cometary nuclei spend most of their lives unseen in the outer reaches of their orbits, far from the Sun, but gravitational disturbances may send some of them on highly elliptical orbits towards the inner Solar System. As a comet nucleus approaches the Sun, the dirty snowball warms up and releases gas and dust to form a coma, perhaps 10 times the diameter of Earth, yet so tenuous that it is transparent. Gas and dust stream away from the coma to produce tails that, in extreme cases, could stretch from Earth to the Sun. Comets have two tails, one of dust and one of gas. Gas tails consist of ionized molecules and have a bluish colour; they are almost straight and are carried directly away from the Sun by the charged particles of the solar wind. Dust tails are curved because the dust particles lag behind the comet's motion; they appear yellowish because the particles reflect sunlight. Cometary dust disperses into space where it is eventually swept up by the planets or falls into the Sun. Dust particles from comets produce the bright streaks known as meteors when they burn up in Earth's atmosphere.

➧ *Asteroids, p109; Meteorites, Meteors and Meteoroids, p111*

NUCLEUS

Generally, the inner part of an object. 1. In a comet, the nucleus is the solid body, composed mainly of water ice, which lies within the coma. 2. In a spiral or barred spiral galaxy, the nucleus is the concentration of material (stars, gas and dust) at the centre.

COMA

1. A defect in a telescope caused when light reaching the telescope at an angle to its axis is spread into a fan-shaped image instead of being focused to a point.

2. The globular cloud of dust and gas that forms around the nucleus of a comet as it approaches the Sun.

CLASSIFICATION OF COMETS

Comets are divided into two broad categories, according to their orbital period (time taken to orbit the Sun). Short-period comets, which include the famous Halley's Comet, are defined as those with orbital periods less than 200 years. These come from a region beyond Neptune called the Edgeworth-Kuiper belt (or sometimes simply the Kuiper Belt).

Long-period comets have orbital periods of thousands or even millions of years, and are thought to originate in a region named the Oort cloud, a hypothetical swarm of cometary nuclei surrounding the Solar System in all directions, roughly halfway to the nearest stars. The gravitational effects of occasional passing stars send some of these dormant comets towards the inner Solar System, where they develop comas and tails.

))**▶** *Edgeworth-Kuiper belt, below; Oort Cloud, below*

PERIODIC COMET

A short-period comet (that is, one with a period less than 200 years) that has been observed to orbit the Sun more than once.

The official designation of a periodic comet is prefixed with 'P/', as in 'P/Halley'. Periodic comets 'wear out' as they lose gas and dust with repeated passages around the Sun.

EDGEWORTH-KUIPER BELT

A region beyond the orbit of Neptune containing KBOs (Kuiper Belt Objects), pieces of debris consisting of a mixture of rock and ice. Short-period comets are thought to originate from the belt. The planet Pluto is sometimes regarded as the largest member of the belt. Triton, the largest satellite of Neptune, was probably once a KBO.

))**▶** *Oort Cloud, below*

OORT CLOUD

A reservoir of icy debris surrounding the entire Solar System from which long-period comets are thought to originate. Long-period comets are those that have orbital periods of thousands or even millions of

years. Although the Oort cloud has yet to be observed, it is believed to surround the Solar System at a distance of 100,000 times Earth's distance from the Sun – roughly halfway to the nearest star. Gravitational disturbances by neighbouring stars occasionally put a dormant comet into an orbit that takes it into the inner solar system. As it approaches the Sun it develops a coma and a gas tail and dust tail.

COMET SHOEMAKER–LEVY 9

Comet that collided with the giant planet Jupiter in July 1994. Carolyn and Eugene Shoemaker and their colleague David Levy had discovered the comet in 1993. Calculations showed that it had been in orbit around Jupiter for over 60 years and after a close approach in 1992 had broken into a chain of more than 20 fragments described as a 'string of pearls'. These hit the planet over a period of a week in July 1994. Dark spots were created in Jupiter's clouds up to 3,000 km (2,000 miles) across. These spots gradually merged into a dusky belt that took over a year to disperse. If the fragments of Comet Shoemaker-Levy 9 had fallen on Earth they would have blasted out craters 60 km (37 miles) across and caused significant climatic changes.

))**▶** *Near Earth Asteroids, p110*

METEORITES, METEORS AND METEOROIDS

Most interplanetary matter is in the form of fine dust particles, termed 'meteoroids'. If these encounter Earth, they either burn up high in the atmosphere to produce a 'shooting star' (a meteor) or, in the case of microscopic particles, settle gently through the atmosphere to Earth's surface as 'micrometeorites'. An object heavier than about 1 g (c. 0.04 oz) can survive its fiery passage and reach the surface of the Earth. The fragments are called 'meteorites'. Orbits of certain incoming meteorites have been calculated from eyewitness descriptions and photographs of their passage through the atmosphere. All turn out to have come from the asteroid belt – they are fragments of asteroids.

Almost all meteorites are 4.5 billion years old and hence date back to the formation of the Solar System. But a handful are younger than this. Some, which contain the same minerals as the surface of the Moon, have ages

between 4.0 and 2.8 billion years; these were ejected from the Moon by impacts. A group of meteorites of volcanic composition, all but one of which have ages of 1.3 billion years or less, contain bubbles of gas which exactly match the composition of the Martian atmosphere. These gases became trapped within the rocks when they were blasted off the surface of Mars by an impact; the rocks subsequently orbited the Sun for a few million years before finally encountering Earth.

A meteorite weighing more than a few hundred tonnes forms a crater when it hits Earth. A crater-forming impact happens on Earth every 5,000 years or so, causing localized destruction. The very largest impacts have global

ABOVE: Meteor Crater, near Flagstaff, Arizona, USA. The crater was formed around 50,000 years ago when an iron meteorite struck the ground creating a hole 1.2 km (three-quarters of a mile) in diameter.

consequences, chiefly because they throw a dust cloud into the upper atmosphere that can envelop Earth, blocking out sunlight for years. Events such as this appear to have happened a few times during Earth's lifetime. One huge impact 65 million years ago may have contributed to the extinction of the dinosaurs.

)))▶ *Death of theDinosaurs, p113; Classification of Meteorites, below*

CLASSIFICATION OF METEORITES

Meteorites are divided into stones, irons and stony-irons, on the basis of their composition. Stony meteorites are subdivided into chondrites and achondrites. Chondrites, the most common meteorites of all, contain chondrules, rounded objects 1 mm (¹⁄₂₅ in) or so in size, which were once suddenly melted and then rapidly cooled, which could have happened either in the

calculated over all wavelengths.

OPACITY

A measure of the ability of a body to absorb or scatter radiation.

CHANDRASEKHAR LIMIT

The maximum permitted mass of a white dwarf star if it is to avoid collapsing further to become a neutron star or black hole. The limit (around 1.4 solar masses) depends on the white dwarf's composition and rotation rate. Calculated by Subrahmanyan Chandrasekhar (1910–95).

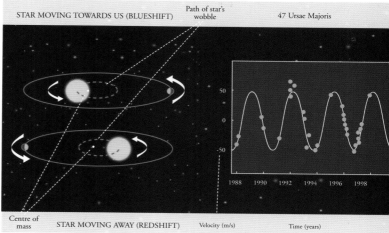

STAR MOVING TOWARDS US (BLUESHIFT) Path of star's wobble   47 Ursae Majoris

50
0
-50

1988 1990 1992 1994 1996 1998

Centre of mass STAR MOVING AWAY (REDSHIFT) Velocity (m/s) Time (years)

ABOVE: Measuring the radial velocity of a star can reveal the presence of an orbiting planet. It can be detected by measuring the 'blueshift' or 'redshift' of the star's light. Here the dots show measured data of 47 Ursa Majoris, the line is the sinewave that best fits the data. Calculations from this graph have revealed the existence of a planet that takes about three years to orbit its star.

EQUIPOTENTIAL SURFACE

A boundary around a celestial body or system at which the gravitational field is constant. In a close binary system the equipotential surface forms two regions called Roche lobes, each of which contains one of the stars. Material from either star can pass through either of two Lagrangian points. Mass transfer takes place at the inner Lagrangian point, while mass is lost from the system at the outer Lagrangian point.

PROPER MOTION

The apparent angular motion of a star on the celestial sphere resulting from its transverse motion (motion across the line of sight) relative to the Solar System. Barnard's star has the greatest known proper motion of 10.3 arcseconds per year.

))》 *Barnard's Star, p139*

RADIAL VELOCITY

The component of velocity of an object in the line of sight, either towards or away from the observer. The radial velocity can be calculated from the Doppler shift of spectral lines.

WIEN'S DISPLACEMENT LAW

Stars, planets and other astronomical bodies can be good approximations to black bodies. One of the characteristics of black-body radiation is that the wavelength of peak emission depends on the temperature of the body.

The German physicist Wilhelm Wien (1864–1928) studied this relationship and, in 1893, came up with an expression which is now known as Wien's displacement law. It relates the wavelength of the peak of the spectrum to the temperature of the body:

wavelength of peak emission (λ_{max}) = 0.0029/T – where λ_{max} is the wavelength of peak emission in metres and T is the temperature in degrees Kelvin. This allows the temperatures of celestial bodies to be estimated.

))》 *Spectra, p118*

ANGULAR MOMENTUM

A measure of the quantity of rotational motion contained in a body or a system of masses. Conservation of angular momentum means that if a collapsing cloud is rotating, it will spin faster and faster as it shrinks. Most stars, particularly lower-mass ones like the Sun, spin rather slowly, having lost most of their angular momentum. For these cooler stars, magnetic fields connect the body of the star to material spun off the surface, enabling the material to carry off the star's angular momentum. More massive stars do not have such magnetic assistance and most continue to spin rapidly.

STELLAR PHENOMENA

LEFT: As a beam of white light enters a prism, the beam bends and fans out into different colours. It is bent again on leaving the prism, so creating the familiar band of colours known as a spectrum.
BELOW: A hot opaque object, such as a star, emits a continuous spectrum. A hot, thin gas emits light at certain wavelengths, forming an emission line spectrum. Where a cool, thin gas lies between the star and the observer, it absorbs light from the star, forming an absorption line spectrum.

spectrum may itself be observed as a reflection spectrum in the light reflected from clouds.

Each chemical element produces its own characteristic set of lines, which may appear either in emission or absorption, thereby allowing analysis of the composition of celestial objects. The measured wavelengths of emission and absorption lines in astronomical objects often differ from the wavelengths of the same elements in the laboratory. This difference is due to the fact that the gas in which the lines arise is moving with respect to the observer because space is expanding, and the wavelengths of the lines are being changed by the Doppler shift.

The Doppler shift also affects the width of the spectral lines. The atoms in a gas are all moving at different speeds, and in consequence the emissions from each atom will have slightly different wavelengths. The result is that the line is slightly broadened, and the lines from a hot gas are broader than the lines from a cold gas, an effect known as Doppler broadening. The line width tells astronomers the temperature of the gas. Doppler broadening can also

SPECTRA

The band of colour produced by shining white light through a prism is known as a spectrum. It is caused by different wavelengths of light travelling at different speeds in the glass and being refracted by different angles at the surfaces. A similar effect is achieved by a diffraction grating, a simple device that splits white light into several coloured beams. Either can be used as the basis of instruments to form and study the spectra of astronomical objects. The general appearance of the spectrum depends upon the physical conditions in the source of light and any absorbing matter along the line of sight.

Several different types of spectrum may be distinguished according to the conditions in which the radiation is emitted. These include continuous spectra, absorption spectra, emission line spectra and reflection spectra. More than one type of spectrum may be present at the same time. The spectrum of the Sun, for example, consists of a continuous spectrum together with bright emission lines and dark absorption lines. The entire solar

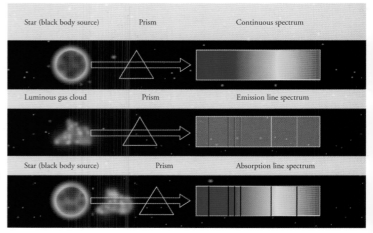

interstellar medium, this gas can later be swept up into new generations of stars. These new stars will have an enriched chemical composition and will evolve slightly differently from their predecessors. The heavy elements made in a supernova explosion are essential for the creation of rocky planets – and of life.

There are two principal types of supernovae with distinctly different origins distinguished on the basis of features in its spectrum. After the explosion of a supernova, the core remains as a neutron star or if it is over about 3 solar masses, a stellar black hole, embedded in a supernova remnant.

)))⬧ *Spectra, p118; Pulsar, p128; Type I Supernova, Type II Supernova, below*

TYPE I SUPERNOVA

Type I supernovae lack hydrogen in their spectra as the stars have no hydrogen left when they explode. Type Ia is thought to arise from the detonation of a white dwarf in a close binary system when gas from the companion spills on to the white dwarf as the companion expands towards the end of its own life. This extra mass compresses and heats the white dwarf, igniting

the carbon and oxygen in its core in a nuclear explosion. Subtypes known as Ib and Ic apparently result from the explosion of single massive stars that have lost their outer layers, probably Wolf-Rayet stars.

)))⬧ *Type II Supernova, below*

TYPE II SUPERNOVA

Type II supernovae are the explosion of high-mass stars at the ends of their lives. They show the existence of hydrogen in their spectra, from the ejected outer layers of the star.

SUPERNOVA REMNANT

Material left over from the explosion of a supernova. These are a type of emission nebula that are heated either by interaction with the interstellar medium or by radiation from the pulsar formed in the supernova explosion.

)))⬧ *Emission Nebula, p57; Pulsar, p128; Interstellar Medium, p149*

ABOVE: A Type 1a supernova (the bright spot at lower left) in the outskirts of galaxy NGC 4526.

ABOVE: A Hubble Space Telescope view of part of the Cygnus Loop, the remains of a star that exploded as a supernova some 30,000 years ago.

NEUTRON STARS

If a dying star retains too much mass to qualify as a white dwarf it will collapse further, into a neutron star. Under extreme gravitational pressure, electrons combine with protons to form the chargeless particles called neutrons. The bulk of the star becomes a 'sea' of free neutrons with a density over a million times greater than that of a white dwarf, topped by a solid crust consisting of a latticework of atomic nuclei.

Neutron stars are probably left behind by many supernovae, although if their weight-loss routine is insufficient to take them below about three solar masses then further collapse awaits, into the oblivion of a black hole. Some neutron stars may arise from the addition of gas to a white dwarf from a companion star in a binary system, tipping it over the Chandrasekhar limit.

Neutron stars were predicted theoretically in the 1930s but the first was not discovered until 1967, when radio astronomers detected rapidly pulsating sources that were dubbed pulsars. Neutron stars have since been found in binary systems where gas from a companion falls on to the neutron star, heating up and emitting X-rays and at the centres of supernova remnants.

))➤ *Pulsar, below; Black Holes, p54; Chandrasekhar Limit, p117; Supernovae, p126; Supernova Remnant, p127*

PULSAR

A rapidly rotating neutron star detected by pulses of radiation, usually at radio wavelengths. Neutron stars have strong magnetic fields, because the magnetic field of their parent star became concentrated when it collapsed.

Charged atomic particles race along the magnetic field lines, beaming high-energy radiation in their direction of travel. Since the field is particularly strong at the magnetic poles, the beam shines out most intensely from there. And because the magnetic axis is usually not aligned with the rotation axis, the beam sweeps around like a lighthouse as the star spins, from once every several seconds to hundreds of times a second depending on the pulsar concerned. The powerful radiation from a pulsar spans the entire electromagnetic spectrum; it is powered by the rotational energy of the star – so, over thousands of years, the pulsar gradually slows down. However, this spin-down is not always smooth; occasional sharp increases, called glitches, are recorded, apparently due to 'starquakes' in the pulsar's crust.

STELLAR BLACK HOLE

If the dying core of a star has more than about three solar masses, it has enough gravity to overcome the neutron degeneracy pressure (the immense pressure exerted by close-packed neutrons) within a neutron star and it becomes a stellar black hole. Such black holes are difficult to detect unless they are in binary systems accompanied by a normal star when an accretion disk of hot gas from the normal star accumulates around the black hole, creating a powerful X-ray source. Accretion disks can also form around neutron stars in binaries, so we cannot assume that every X-ray-emitting binary contains a black hole. In most known binary X-ray sources the compact object has a mass about that of the Sun, and so is probably a neutron star, but in a few cases the mass is much greater than three solar masses, beyond the theoretical limit for a neutron star. These must, it seems, be black holes.

))➤ *Neutron Stars, above; Black Holes, p54; X-Rays, p59; Accretion Disk, p146*

BINARY STARS

Roughly half of all the stars in the sky are found to be in binary systems, in which two companion stars orbit their common centre of mass with periods ranging from hours to many thousands of years. In keeping with Kepler's laws, the wider the separation between the stars, the longer their orbital period. Binary stars may be born together from a rapidly rotating protostar that split, or one may capture the other in a close encounter after birth. Some may eclipse each other, causing variability in their observed brightness.

Binary stars are generally classified according to the means by which their binary nature is observed: in astrometric binaries only one component is observed, but the companion is inferred by perturbations in its proper motion; spectroscopic binaries are revealed by the Doppler shift of lines in their spectrum; in visual binaries both components can be resolved through a telescope. Binary stars are important for determining

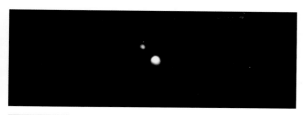

ABOVE: The yellow and blue components of the double star Albireo.

information about stellar masses. Some binary stars are so close that transfer of material can occur between the components affecting their evolution and producing variations in their light output.

**⫸ *Double Stars, below; Johannes Kepler, p27;
Doppler Effect, p60***

DOUBLE STARS

Two stars that appear close together in the sky. Optical doubles are stars that appear to be close due to the line-of-sight effect while physical doubles are gravitationally bound.

VISUAL BINARIES

If both the components of a binary star are far enough apart as seen from Earth then it is known as a visual binary. The larger the telescope's aperture, the closer the stars that can be distinguished, although the ease of doing this will also depend on the relative brightness of the pair: a bright star may drown out the light of a feebler companion. Some apparent double stars are not connected and are simply viewed along a similar line of sight; these are called visual doubles, but they are much less common than true binaries. If a true binary is observed for long enough, signs of orbital motion will be detected, as the stars gradually change their separation and orientation relative to each other; careful measurement of these quantities can be used to determine the masses of the stars.

RIGHT: The sequence of changing spectral lines reveals the presence of a spectroscopic binary. When both stars move across the observers line of sight, the lines merge (1 and 3). When one star is approaching and the other is receding, the lines have become separated (2 and 4).

BINARY PULSARS

In 1974 two US astronomers discovered a pulsar in orbit around a neutron star. By making careful measurements of the pulse times, they could calculate the orbits of the stars with great precision. General relativity predicts that the two stars, orbiting each other in less than eight hours, will be radiating energy in the form of gravitational waves, and this should be revealed by the orbits slowly contracting as the stars spiral in towards each other with increasing speed. Within a few years, the orbits were found to be shrinking at precisely the rate predicted by general relativity. So compelling was this evidence for gravitational waves that the two astronomers, Joseph Taylor (b. 1941) and Russell Hulse (b. 1950), were deservedly awarded the Nobel Prize for physics in 1993.

⫸ *Gravitational Waves, p61; Neutron Stars, Pulsar, p128*

SPECTROSCOPIC BINARIES

Many stars are too close together to be resolved directly through a telescope, but they can be detected by spectroscopy. If the lines in a star's spectrum are moving cyclically back and forth in wavelength, it could be due to the Doppler shift as the star moves in orbit around its centre of mass with a companion (although it could be a single star that is pulsating in size). If the spectrum of only one component can be seen it is termed a single-line spectroscopic binary, but when

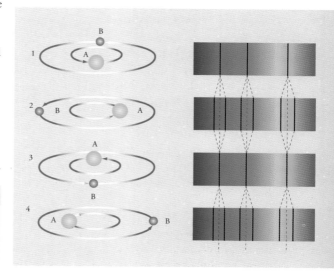

the two are of comparable brightness a double-line spectroscopic binary may be observed. By monitoring such a spectroscopic binary throughout its orbital cycle, astronomers can calculate the orbital elements, quantities which describe the characteristics of the orbit: its period, shape, size and orientation in space, and the speed of the stars. This is one of the techniques used to discover many otherwise invisible planetary companions of stars.

ECLIPSING BINARIES

The orbits of some spectroscopic binaries are orientated side-on to us, so that one star periodically passes in front of the other, causing an eclipse and reducing the total light we receive from the binary for a while. Such a system is termed an eclipsing binary. A light curve (a graph of brightness versus time) can be plotted of the brightness variations caused by the eclipses which gives valuable information about the binary, including the shape of the stars, for it turns out that they are not always spherical: when the components are close together, they can deform each other into egg shapes which leads to additional variability as the observer sees the star either side-on or end-on. In the closest pairings the stars seem to be touching: these are termed contact binaries.

CATACLYSMIC BINARIES

Cataclysmic binaries (or variables) are binary stars in which the components are very close to each other. In such systems, a cool red dwarf star is partnered by a white dwarf. The two are less than one million km (600,000 miles) apart and orbit with periods less than half a day. Gas is drawn off the red dwarf by the gravitational strength of the white dwarf and it spirals on to an accretion disk around the compact star creating a bright hot spot at the point where the infalling stream of gas hits the disk. Every few days or weeks, instabilities in the disk cause surges in brightness of as much as 250 times (six magnitudes). The result is termed a dwarf nova.

Gas eventually spirals down through the accretion disk on to the surface of the white dwarf. When sufficient gas has built up, a 'hydrogen flash' can occur, a nuclear

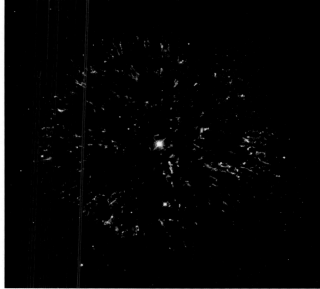

explosion which blows off a thin outer layer, and the star's brightness flares up 25,000 times (11 magnitudes) or so in a nova explosion. Nova eruptions may repeat after hundreds or thousands of years, and several such recurrent novae are known. In extreme cases, where the composition of the white dwarf is just right, the nuclear detonation may be so severe as to trigger a supernova of Type Ia.

)))▶ *Type I Supernova, p127; Binary Stars, p128; Accretion Disk, p146*

NOVAE

The name 'nova', Latin for 'new', was given in ancient times when astronomers saw stars appear where none had been seen before and concluded that they were indeed new arrivals in the firmament. They are now known to be cataclysmic binaries. Dozens of novae are thought to occur in our Galaxy each year, although only occasionally does one become bright enough to be visible to the naked eye. A typical nova brightens by 10 magnitudes (a factor of 10,000)

ABOVE: Ejected gas surrounding Nova Persei, which exploded in 1901. At maximum the nova was around 200,000 times as luminous as the Sun. A nova is a type of cataclysmic binary.

or more in a few days and then declines slowly back to its original luminosity.

Am STARS

Stars which show an excess of most heavy elements but are strangely deficient in some, notably calcium and scandium, are known as Am stars. Such stars must rotate slowly to avoid stirring up the atmosphere, and most seem to have achieved this through the braking effect of tidal forces supplied by a companion star.

))))▶ *Ap Stars, below*

Ap STARS

A group of stars known as peculiar A stars (or Ap stars) have magnetic fields thousands of times stronger than the Sun's. As a result, certain elements such as silicon, chromium and europium are concentrated in spots on their surfaces, so that the stars' spectral features vary as the star rotates.

BARIUM STARS

Barium stars are giant stars that had the misfortune to be too closely partnered to a star that completed its evolution and shot out the products of its nuclear burning all over its companion, especially barium. They therefore show enhancements of elements such as barium and strontium in their spectra.

MAIN SEQUENCE STARS

Stars spend most of their lives on a region of the Hertzsprung-Russell diagram known as the main sequence. Here they 'burn' their initially plentiful hydrogen fuel by nuclear reactions in their core. During this time they will maintain a steady surface temperature and luminosity. The star's position on the main sequence and how long it will spend there depends on its mass. The most massive stars (up to about 100 times the mass of the Sun) populate the upper end of the main sequence and the least massive stars, the lower end. Stars on the main sequence are termed dwarfs even though the largest of them are 15 times bigger than the Sun.

))))▶ *Hertzsprung-Russell Diagram, p115*

CEPHEIDS

An important category of pulsating stars are the Cepheids, named after their prototype Delta [δ] Cephei. These are yellow supergiants of spectral type F and G which oscillate on anything from a daily to monthly basis. As they do so their visible light changes as much as sixfold (two magnitudes), their temperature by over a thousand degrees and their diameter by about 10 per cent. Cepheids are highly luminous and so can be seen at great distances – even in other galaxies – and as their luminosity is related to their period of oscillation (the brightest ones having the longest periods), this makes them valuable as distance indicators despite their relative scarcity. If a star can be identified as a Cepheid, its absolute magnitude can be determined by measuring its period, and its distance can then be deduced.

))))▶ *Period-Luminosity Relation, p116*

RR LYRAE STARS

A type of pulsating variable star that varies over about 0.2 to 2 magnitudes in a period of less than one day, generally found in globular clusters. Fainter than Cepheids, they are still valuable as distance indicators. They have a similar range of brightness variation to the Cepheids but much shorter periods, no more than about a day. In the HR diagram they are located on the horizontal branch between the main sequence and the red giants at spectral types A and F, and are stars of low mass, burning helium in their cores, following the helium flash.

))))▶ *Low-Mass Stars, p124*

MIRA VARIABLES

Long-period variables or Mira stars (named after their prototype, Mira or Omicron [o] Ceti) are pulsating cool red giants or supergiants. They change by up to 11 magnitudes, being about two or three times larger when at their biggest than at their smallest. During the pulsation cycle their surface temperatures range from about 1,600°C (2,900°F) to 2,300°C (4,200°F) – that is, if their tenuous outer layers can be thought of as a surface, since they have densities lower than a vacuum produced in an Earth-bound laboratory. It is mainly the change in temperature rather than size that is responsible for the variations in the visible light from Miras; at minimum, much of their energy is emitted in the invisible infrared. Another feature of this phase of evolution is extensive mass loss in a 'wind' of particles from the star's surface; half the star's mass, or more, can seep into space in this fashion. Equally significantly, huge convection cells in the star's outer layers can dredge carbon from the star's core to the surface. Carbon is blown off by the stellar wind and condenses into solid dust grains in interstellar space.

)))➤ *Red Giant, below; Supergiant, p133*

R CORONAE BOREALIS STARS

Among the stranger types of variable, the R Coronae Borealis stars stand out. These are supergiants of spectral types F or G, rich in carbon. Atmospheric instabilities that occur every few years cause R Coronae Borealis stars to eject great clouds of carbon into space, where it immediately turns to soot and blocks the light from the star.

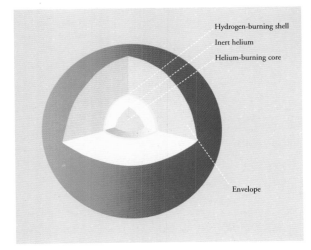

Hydrogen-burning shell

Inert helium

Helium-burning core

Envelope

ABOVE: By the time a star becomes a red giant, the nuclear fusion of hydrogen over billions of years has already made a core of burning helium, contained within a helium shell. Surrounding this is a further shell of burning hydrogen, beyond which is the tenuous envelope.

RED GIANT

A star which has finished burning hydrogen in its core and which is expanding due to burning hydrogen in a series of shells. In response to this, the

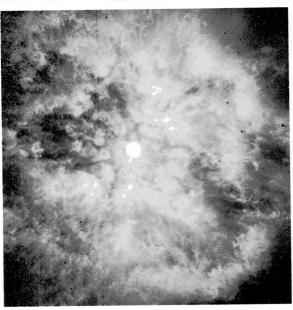

BELOW: A Wolf-Rayet star, WR124 in Sagittarius, surrounded by a cloud of hot gas being ejected in a furious stellar wind.

star's outer layers expand and cool, making it appear red in colour. The swelling star moves off the main sequence and moves first to the right on the Hertzsprung-Russell diagram as its surface temperature drops and then upwards to reflect its increasing luminosity. Red giants have diameters 10 to 1,000 times that of the Sun with surface temperatures as low as 3,600°C (6,500°F). No energy is now being generated in the helium-filled core, which begins to contract and heat up. Stars of two solar masses reach this turning point in their evolution less than two billion years after birth, while for the smallest, the timescale is longer than the present age of the Universe, so none of those have yet left the main sequence.

WOLF-RAYET STARS

High-mass stars leaving the main sequence suffer huge stellar winds with speeds up to 2,000 km/s (1,200 mi/s). These winds strips off their outer hydrogen layers at a rate of 10 million million tonnes per second. With the outer covering removed, the products of the nuclear burning in the star's interior come into view, giving rise to strong, broad emission lines in its spectrum, which originate in a dense expanding envelope.

These stars, often found at the centres of planetary nebulae, are known as Wolf-Rayet stars after their French discoverers, Charles Wolf (1827–1918) and Georges Rayet (1839–1906). Those showing prominent nitrogen lines are designated WN stars. Later in the star's evolution, when the helium produced in the CNO reactions is utilized to produce carbon (and the mass loss continues), the nitrogen lines disappear to be replaced by strong carbon and oxygen lines; these are classified as WC stars.

HYPERGIANTS

The hypergiants are among the most luminous stars in the Galaxy, perhaps 100,000 times brighter than the Sun in the visual part of the electromagnetic spectrum. They have masses of about 30–100 Suns. They have spectral types ranging from B to M and have been allocated the luminosity class Ia+ or even 0, setting them above the brightest normal supergiants. Hypergiants often move around the Hertzsprung-Russell diagram even within a human lifetime, these excursions being accompanied by sporadic events of mass-loss which generate shells of material around the star. A well-known example, Rho Cassiopeiae, is to be found near the familiar W-shape of stars in Cassiopeia.

)))➤ *Hertzsprung-Russell Diagram, p115*

SUPERGIANT

The largest and most luminous stars known, with spectral types from O to M. Red (M-type) supergiants have the largest radii, of the order of 1,000 times that of the Sun. Supergiants lie above the main sequence and giant

BELOW: The vast Eta Carinae nebula is a glowing gas cloud that appears four times wider than the Moon, and contains in its brightest portion, just above centre, the peculiar eruptive luminous blue variable star Eta Carinae.

region on the Hertzsprung-Russell diagram and are often variable, due to instabilities created by radiation pressure.

LUMINOUS BLUE VARIABLES

The most massive stars, of spectral type O, are forever losing huge quantities of gas from their surfaces, but when they leave the main sequence matters become far worse and they become luminous blue variables (LBVs). Violent instabilities set in, resulting in episodes of huge mass loss, probably every few hundred or thousand years, which cause the star to brighten by several magnitudes, and for its spectral type to change as its outer envelope of gas lifts off into space. The expanding shell will produce striking emission lines in the spectrum, with adjacent absorption lines where the shell absorbs light from the star below. Such dual structures are known as P Cygni lines since they were first identified in the spectrum of one of the most famous LBVs, the high-luminosity blue supergiant P Cygni.

)))➤ *Eta Carinae, p135*

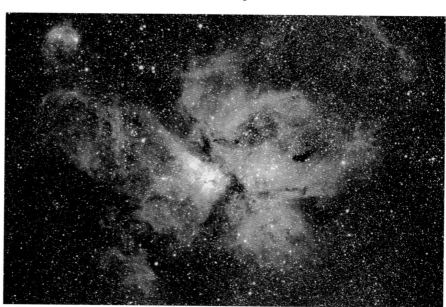

CONSTELLATIONS AND STARS

The sky is divided into 88 sections known as constellations, a combination of the 48 figures known to the ancient Greeks and others introduced more recently. Their names and boundaries were officially laid down by the International Astronomical Union in 1930. Constellations originated in the distant past as easy-to-remember patterns of stars, but nowadays astronomers regard them merely as areas of sky convenient for locating and naming celestial objects. The areas are bounded by arcs of right ascension and declination (epoch 1875). The patterns on which constellations were originally based are purely the product of human imagination – with very few exceptions, the stars in a constellation are unrelated and lie at widely differing distances from us.

)))⟩ *Asterisms, below*

ASTERISMS

Asterisms are unofficial patterns that can consist of stars from one or more constellation. The Square of Pegasus is an example of an asterism composed of stars from two constellations (Pegasus and Andromeda). The Plough, the Sickle of Leo, and the Teapot of Sagittarius are all asterisms within a given constellation.

NAMES OF STARS

Stars bear a variety of names, some traditional, others named after astronomers who discovered or studied them. All bright stars are identified by either a letter (usually Greek) or a number, along with the name of the constellation that contains them. For example, Sirius is also Alpha [α] Canis Majoris, – the genitive (possessive) case of the constellation name is always used in this context. In 1603 Johann Bayer (1572–1625) published an atlas in which the stars were identified by Greek letters, and so these are usually termed Bayer letters. A supplementary system identifies stars by numbers – called Flamsteed numbers, after the first Astronomer Royal of England. An example is 61 Cygni. Variable stars, if they are not already identified on one of the preceding systems, have a different style of nomenclature. Names of the brighter ones consist of one or two Roman letters, e.g. P Cygni, VV Cephei. Others are called V (for variable) with a three- or four-digit number.

BELOW: The Andromeda Galaxy is only just visible to the naked eye.

ANDROMEDA

Andromeda contains M31, the Andromeda Galaxy, the most distant object you can glimpse with the naked eye. A telescope is needed to spot its small, elliptical companion galaxies M32 and M110. Andromeda also contains Gamma [γ] Andromedae, a binary of contrasting colours.

AQUARIUS

The zodiacal constellation of Aquarius the Water-bearer is the home of two famous planetary nebulae: the Helix and the Saturn. It also contains a group of four stars arranged in a Y shape, forming an attractive sight through binoculars. These stars